Undergraduate Texts in Mathematics

Editors

S. Axler
F.W. Gehring
K.A. Ribet

D0902348

Springer
New York
Berlin
Heidelberg
Barcelona
Hong Kong
London
Milan
Paris
Singapore
Tokyo

Undergraduate Texts in Mathematics

Abbott: Understanding Analysis.

Anglin: Mathematics: A Concise History and Philosophy.
Readings in Mathematics.

Anglin/Lambek: The Heritage of Thales.
Readings in Mathematics.

Apostol: Introduction to Analytic Number Theory. Second edition.

Armstrong: Basic Topology.

Armstrong: Groups and Symmetry.

Axler: Linear Algebra Done Right. Second edition.

Beardon: Limits: A New Approach to Real Analysis.

Bak/Newman: Complex Analysis. Second edition.

Banchoff/Wermer: Linear Algebra Through Geometry. Second edition.

Berberian: A First Course in Real Analysis.

Bix: Conics and Cubics: A Concrete Introduction to Algebraic Curves.

Brémaud: An Introduction to Probabilistic Modeling.

Bressoud: Factorization and Primality Testing.

Bressoud: Second Year Calculus.
Readings in Mathematics.

Brickman: Mathematical Introduction to Linear Programming and Game Theory.

Browder: Mathematical Analysis: An Introduction.

Buchmann: Introduction to Cryptography.

Buskes/van Rooij: Topological Spaces: From Distance to Neighborhood.

Callahan: The Geometry of Spacetime: An Introduction to Special and General Relavitity.

Carter/van Brunt: The Lebesgue–Stieltjes Integral: A Practical Introduction.

Cederberg: A Course in Modern Geometries. Second edition.

Childs: A Concrete Introduction to Higher Algebra. Second edition.

Chung: Elementary Probability Theory with Stochastic Processes. Third edition.

Cox/Little/O'Shea: Ideals, Varieties, and Algorithms. Second edition.

Croom: Basic Concepts of Algebraic Topology.

Curtis: Linear Algebra: An Introductory Approach. Fourth edition.

Devlin: The Joy of Sets: Fundamentals of Contemporary Set Theory. Second edition.

Dixmier: General Topology.

Driver: Why Math?

Ebbinghaus/Flum/Thomas: Mathematical Logic. Second edition.

Edgar: Measure, Topology, and Fractal Geometry.

Elaydi: An Introduction to Difference Equations. Second edition.

Exner: An Accompaniment to Higher Mathematics.

Exner: Inside Calculus.

Fine/Rosenberger: The Fundamental Theory of Algebra.

Fischer: Intermediate Real Analysis.

Flanigan/Kazdan: Calculus Two: Linear and Nonlinear Functions. Second edition.

Fleming: Functions of Several Variables. Second edition.

Foulds: Combinatorial Optimization for Undergraduates.

Foulds: Optimization Techniques: An Introduction.

Franklin: Methods of Mathematical Economics.

Frazier: An Introduction to Wavelets Through Linear Algebra.

Gamelin: Complex Analysis.

Gordon: Discrete Probability.

Hairer/Wanner: Analysis by Its History.
Readings in Mathematics.

(continued after index)

Karen Saxe

Beginning Functional Analysis

 Springer

Karen Saxe
Mathematics Department
Macalester College
St. Paul, MN 55105
USA

Mathematics Subject Classification (2000): 46-01

Library of Congress Cataloging-in-Publication Data
Saxe, Karen.
      Beginning functional analysis / Karen Saxe.
         p. cm. — (Undergraduate texts in mathematics)
      Includes bibliographical references and index.

      1. Functional analysis.   I. Title.   II. Series.
   QA320 .S28 2001
   515′.7—dc21   00-067916

Printed on acid-free paper.

Production managed by A. Orrantia; manufacturing supervised by Erica Bresler.

Printed in the United States of America.

9 8 7 6 5 4 3 2 1

ISBN 978-1-4419-2914-3

Springer-Verlag   New York Berlin Heidelberg
A member of BertelsmannSpringer Science+Business Media GmbH

Preface

This book is designed as a text for a first course on functional analysis for advanced undergraduates or for beginning graduate students. It can be used in the undergraduate curriculum for an honors seminar, or for a "capstone" course. It can also be used for self-study or independent study. The course prerequisites are few, but a certain degree of mathematical sophistication is required.

A reader must have had the equivalent of a first real analysis course, as might be taught using [25] or [109], and a first linear algebra course. Knowledge of the Lebesgue integral is not a prerequisite. Throughout the book we use elementary facts about the complex numbers; these are gathered in Appendix A. In one specific place (Section 5.3) we require a few properties of analytic functions. These are usually taught in the first half of an undergraduate complex analysis course. Because we want this book to be accessible to students who have not taken a course on complex function theory, a complete description of the needed results is given. However, we do not prove these results.

My primary goal was to write a book for students that would introduce them to the beautiful field of functional analysis. I wanted to write a succinct book that gets to interesting results in a minimal amount of time. I also wanted it to have the following features:

- It can be read by students who have had *only* first courses in linear algebra and real analysis, and it ties together material from these two courses. In particular, it can be used to introduce material to undergraduates normally first seen in graduate courses.
- Reading the book *does not* require familiarity with Lebesgue integration.

- It contains information about the historical development of the material and biographical information of key developers of the theories.
- It contains many exercises, of varying difficulty.
- It includes ideas for individual student projects and presentations.

What really makes this book different from many other excellent books on the subject are:

- The choice of topics.
- The level of the target audience.
- The ideas offered for student projects (as outlined in Chapter 6).
- The inclusion of biographical and historical information.

How to use this book

The organization of the book offers flexibility. I like to have my students present material in class. The material that they present ranges in difficulty from "short" exercises, to proofs of standard theorems, to introductions to subjects that lie outside the scope of the main body of such a course.

- Chapters 1 through 5 serve as the core of the course. The first two chapters introduce metric spaces, normed spaces, and inner product spaces and their topology. The third chapter is on Lebesgue integration, motivated by probability theory. Aside from the material on probability, the Lebesgue theory offered here is only what is deemed necessary for its use in functional analysis. Fourier analysis in Hilbert space is the subject of the fourth chapter, which draws connections between the first two chapters and the third. The final chapter of this main body of the text introduces the reader to bounded linear operators acting on Banach spaces, Banach algebras, and spectral theory. It is my opinion that every course should end with material that truly challenges the students and leaves them *asking* more questions than perhaps can be answered. The last three sections of Chapter 5, as well as several sections of Chapter 6, are written with this view in mind. I realize the time constraints placed on such a course. In an effort to abbreviate the course, some material of Chapter 3 can be safely omitted. A good course can include only an outline of Chapter 3, and enough proofs and examples to give a flavor for measure theory.
- Chapter 6 consists of seven independent sections. Each time that I have taught this course, I have had the students select topics that they will study individually and teach to the rest of the class. These sections serve as resources for these projects. Each section discusses a topic that is nonstandard in some way. For example, one section gives a proof of the classical Weierstrass approximation theorem and then gives a fairly recent (1980s) proof of Marshall Stone's generalization of Weierstrass's theorem. While there are several proofs of the Stone–Weierstrass theorem, this is the first that does not depend on the classical result. In another section of this chapter, two arguments are given that no function can be continuous at each rational number and discontinuous at each

irrational number. One is the usual Baire category argument; the other is a less well known and more elementary argument due to Volterra. Another section discusses the role of Hilbert spaces in quantum mechanics, with a focus on Heisenberg's uncertainty principle.

- Appendices A and B are very short. They contain material that most students will know before they arrive in the course. However, occasionally, a student appears who has never worked with complex numbers, seen De Morgan's Laws, etc. I find it convenient to have this material in the book. I usually spend the first day or two on this material.

- The biographies are very popular with my students. I assign each student one of these (or other) "key players" in the development of linear analysis. Then, at a subject-appropriate time in the course, I have that one student give (orally) a short biography in class. They really enjoy this aspect of the course, and some end up reading (completely due to their own enthusiasm) a book like Constance Reid's *Hilbert* [104].

Acknowledgments

It is a pleasure to thank several people for their help with this project. My colleagues David Bressoud and Joan Hutchinson gave advice and encouragement early in my writing process. John Schue read an entire preliminary version, and I thank him for his energy and many thoughtful comments. Per Enflo and Tom Kriete have helped substantially with sections of the book, and I remain indebted to each of them for their contributions. Thanks also go to Steve Abbott, Frédéric Gourdeau, Tom Halverson, Michael Neumann, Karen Parshall, and Tom Ransford for valuable conversations on different bits of the content of the book. My husband, Peter Webb, gave emotional support and much needed help with TEX. Our children, Julian, Alexander, and Zoë, were wonderful during the final stages of this book; they played for many long hours by themselves, and also pulled me away from the computer when I really needed a break. Much of the writing of the book, and general peace of mind, was achieved during a sabbatical in the spring of 2000 spent at the University of Virginia. I thank their mathematics department for its hospitality.

This book has evolved from notes for a course that I have taught three times over several years at Macalester. The students in those courses have motivated me with their enthusiasm for the project. I thank them for their apparent eagerness to read my sometimes not-so-polished notes, and for much help in sorting out good exercise sets. It is fair to say that if it weren't for my students, I would not have written this book.

I welcome your comments, suggestions for improvements, and indications of errors.

Karen Saxe
Macalester College
saxe@macalester.edu
April 2001

Contents

Introduction: To the Student

Functional analysis was developed in the last years of the nineteenth century and during the first few decades of the twentieth century. Its development was, in large part, in response to questions arising in the study of differential and integral equations. These equations were of great interest at the time because of the vast effort by many individuals to understand physical phenomena.

The unifying approach of functional analysis is to view functions as points in some abstract vector (linear) space and to study the differential and integral equations relating these points in terms of linear transformations on these spaces. The term "functional analysis" is most often credited to Paul P. Lévy (1886–1971; France).[1] The rise of the field is consistent with a larger move toward generality and unification in mathematics. Indeed, this move can be viewed as part of a more general intellectual trend, and it is interesting to compare it to analogous movements in other fields such as philosophy, music, painting, and psychology.

Maurice Fréchet (1878–1973; France) is usually credited with the first major effort to develop an abstract theory of spaces of functions. Much of this work appears in his 1906 doctoral thesis. Many other names are associated with the birth and development of functional analysis, and you will read about them as you proceed through this text. The works of Stefan Banach (1892–1945; Austria–Hungary, now Hungary) and David Hilbert (1862–1943; Prussia, now Russia) have probably had the greatest influence.

It has been my goal to present the basics of functional analysis in a way that makes them comprehensible to a student who has completed first courses in linear

---

[1] See the biography of Fréchet for more on the origins of this phrase.

algebra and real analysis, and to develop the topics in their historical context. Bits of pertinent history are scattered throughout the text, including brief biographies of some of the central players in the development of functional analysis.

In this book you will read about topics that can be gathered together under the vague heading, "What everyone should know about functional analysis." ("Everyone" certainly includes anyone who wants to study further mathematics, but also includes anyone interested in the mathematical foundations of economics or quantum mechanics.) The first five chapters of the book are devoted to these essential topics. The sixth chapter consists of seven independent sections. Each section contains a topic for further exploration; some of these develop further a topic found in the main body of the text, while some introduce a new topic or application. The topics found in the sixth chapter provide good bases for individual student projects or presentations. Finally, the book concludes with two appendices that offer basic information on, respectively, complex numbers and set theory. Most of the material found in these two sections is not hard, but it is crucial to know before reading the book. The appendices can be read in advance and can be used as reference throughout your reading of the text.

There are plenty of exercises. There is much wisdom in the saying that you must *do* math in order to learn math. The level of difficulty of the exercises is quite variable, so expect some of them to be straightforward and others quite challenging.

There are many excellent books on functional analysis and the other topics that we discuss in this text. The bibliography includes references to classics by the "founding fathers" ([11], [80], [107], for example); some of the standard texts currently used for first-year graduate courses ([44], [47], [111], for example), treatments of historical aspects of our subject ([16], [17], [23], [34], [54], [61], [73], [76], [104], for example); books on related topics ([1], [22], [27], [77], [121], [99], for example); undergraduate real analysis texts ([25], [89], [109], [110], for example); and readable journal articles on topics we discuss ([13], [26], [31], [37], [64], [91], [96], [117], [119], [125], for example).

The list of references is meant to be used, and I hope that you take the opportunity to look at many of the referenced books and articles.

Finally, there is a very good history of mathematics web site run at St. Andrews University:
http://www-groups.dcs.st-and.ac.uk/~history (this address was good as of April 2001).

1

Metric Spaces, Normed Spaces, Inner Product Spaces

The goal of this chapter is to introduce the abstract theory of the spaces that are important in functional analysis and to provide examples of such spaces. These will serve as our examples throughout the rest of the text, and the spaces introduced in the second section of this chapter will be studied in great detail. The abstract spaces—metric spaces, normed spaces, and inner product spaces—are all examples of what are more generally called "topological spaces." These spaces have been given in order of increasing structure. That is, every inner product space is a normed space, and in turn, every normed space is a metric space. It is "easiest," then, to *be* a metric space, but because of the added structure, it is "easiest" to *work* with inner product spaces.

Fréchet developed the general concept of the abstract metric space. The other two types of spaces of interest to us, inner product and normed spaces, are particular types of linear spaces.[1] Giuseppe Peano (1858–1932; Italy) gave the first axiomatic definition of a linear space in 1888 (see [35]). In 1922 Banach wrote down the axioms for a normed space in [10]. The axioms for inner product spaces were presented by John von Neumann (1903–1957; Hungary) [98]. As is most often the way in mathematics, the origins of these works, including anticipation of the axiomatic definitions, can be seen in the work of their predecessors. You will read much more about these sources of motivation throughout the text.

You are probably familiar with the finite-dimensional Euclidean spaces, $\mathbb{R}^n$. As you will see, the spaces of functional analysis are typically infinite-dimensional.

---

[1] A linear space is the same thing as a *vector space*; we will always use the former terminology in order to emphasize the *linearity* that permeates the subject of this book.

The unification of the ideas of linear spaces of finite and infinite dimensions took some time. The publication of two major works in the early 1930s, Banach's *Théorie des Opérations Linéaires* [11] and *Moderne Algebra* by the algebraist Bartel van der Waerden (1903–1996; Netherlands), helped to solidify this unification. At the time that Peano wrote down his axioms for a linear space, differential equations was already an important branch of study. Connections between the fields of differential equations and matrix theory already existed. For example, Lagrange used methods that we would now refer to as "eigenvalue methods" to solve systems of simultaneous differential equations in several variables. But it would be a while until the connection was truly recognized and understood. Joseph Fourier (1768–1830; France) had already been studying countably infinite systems of such equations (see Chapter 4), but his method involved considering the "subsystems" of the first n equations and then letting n tend to infinity.

1.1   Basic Definitions and Theorems

The idea of an inner product space is to describe an abstract structure with the desirable properties of Euclidean space: a distance-measuring device and a way of determining orthogonality. A "metric" is simply a way of measuring distances between points of the space. For example, the space $\mathbb{R}^n$ is a metric space with metric

$$d(x, y) = \sqrt{(x_1 - y_1)^2 + (x_2 - y_2)^2 + \cdots + (x_n - y_n)^2}$$

for $x = (x_1, x_2, \ldots, x_n)$ and $y = (y_1, y_2, \ldots, y_n)$ in $\mathbb{R}^n$. This metric is the standard Euclidean metric on $\mathbb{R}^n$.

In general, a *metric space* (M, d) is defined to be a set M together with a function $d : M \times M \to \mathbb{R}$ called a *metric* satisfying four conditions:

  (i) $d(x, y) \geq 0$ for all $x, y \in M$ (nonnegativity),
  (ii) $d(x, y) = 0$ if and only if $x = y$ (nondegeneracy),
  (iii) $d(x, y) = d(y, x)$ for all $x, y \in M$ (symmetry),
  (iv) $d(x, y) \leq d(x, z) + d(z, y)$ for all $x, y, z \in M$ (triangle inequality).

In addition to $\mathbb{R}^n$, what are some examples of metric spaces?

EXAMPLE 1.   Let $M = \mathbb{C}$, with $d(z, w) = |z - w|$.

EXAMPLE 2.   Let M be any set and define

$$d(x, y) = \begin{cases} 1 & \text{if } x \neq y, \\ 0 & \text{if } x = y. \end{cases}$$

This is called the *discrete metric*.

EXAMPLE 3.   Fix a positive integer n and let M be the set of all ordered n-tuples of 0s and 1s. For x and y in M, define $d(x, y)$ to be the number of places in which x and y differ. For example, with $n = 6$,

$$d(001011, 101001) = 2.$$

Many of the topological notions from $\mathbb{R}$ can be extended to the general metric space setting. For example, a sequence $\{x_n\}_{n=1}^{\infty}$ in a metric space (M, d) is said to *converge* to the element $x \in M$ if for each $\epsilon > 0$ there exists a positive integer N such that $d(x_n, x) < \epsilon$ whenever $n \geq N$. As another example, a function f defined on a metric space (M, d_M) and taking values in another metric space (N, d_N), is *continuous at* $x_0 \in M$ if given any $\epsilon > 0$ there exists a $\delta > 0$ such that

$$d_N\Big(f(x), f(x_0)\Big) < \epsilon \qquad \text{whenever} \qquad d_M(x, x_0) < \delta.$$

We shall discuss more "topology" in the next chapter.

Many of the metrics that we will be interested in arise from "norms." A (real) *normed linear space* $(V, \|\cdot\|)$ is a (real) linear space V together with a function $\|\cdot\| : V \to \mathbb{R}$ called a *norm* satisfying four conditions:

(i) $\|v\| \geq 0$ for all $v \in V$ (nonnegativity),
(ii) $\|v\| = 0$ if and only if $v = 0$ (nondegeneracy),
(iii) $\|\lambda v\| = |\lambda| \cdot \|v\|$ for all $v \in V$ and $\lambda \in \mathbb{R}$ (multiplicativity),
(iv) $\|v + w\| \leq \|v\| + \|w\|$ for all $v, w \in V$ (triangle inequality).

We now give four basic normed linear spaces. More interesting examples will be given in the next section.

EXAMPLE 1. $V = \mathbb{R}$ with $\|x\| = |x|$.

EXAMPLE 2. $V = \mathbb{R}^n$ with $\|x\| = \sqrt{x_1^2 + x_2^2 + \cdots + x_n^2}$ for $x = (x_1, x_2, \ldots, x_n)$. This is the *usual, standard,* or *Euclidean* norm on $\mathbb{R}^n$. It is usually denoted by $\|\cdot\|_2$.

EXAMPLE 3. $V = \mathbb{C}$, with $\|z\| = |z|$.

EXAMPLE 4.   We can define many norms on $\mathbb{R}^n$. For example, both

$$\|x\|_1 = |x_1| + |x_2|$$

and

$$\|x\|_\infty = \max(|x_1|, |x_2|)$$

define norms on $\mathbb{R}^2$. These can be extended in the obvious way to $\mathbb{R}^n$, and we will see later, in the exercises, that they are not really all that different. These norms might seem a bit odd but they are related to the important sequence spaces ℓ^1 and ℓ^∞, which will be defined in the next section.

Theorem 1.1. *Norms always give rise to metrics. Specifically, if $(V, \|\cdot\|)$ is a normed space and $d(v, w)$ is defined by*

$$d(v, w) = \|v - w\|,$$

then d is a metric on V.

PROOF.   Left as Exercise 1.1.5.   □

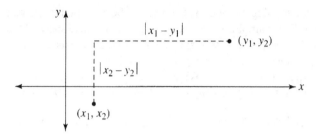

FIGURE 1.1. The metric $d(x, y) = |x_1 - y_1| + |x_2 - y_2|$.

Not all metrics come from norms. For example, the discrete metric cannot come from a norm. This is because $\|\lambda v\| = |\lambda| \cdot \|v\| \to \infty$ as $|\lambda| \to \infty$. Thus, $d(\lambda v, 0) \to \infty$ as $|\lambda| \to \infty$. But $d(\lambda v, 0) = 1$ unless $v = 0$. If we consider the three norms $\| \cdot \|_1$, $\| \cdot \|_2$, and $\| \cdot \|_\infty$ on $\mathbb{R}^2$, only the middle one gives rise to the usual metric on $\mathbb{R}^2$. The other two give rise to other metrics. For example, the metric arising from the first one is

$$d(x, y) = |x_1 - y_1| + |x_2 - y_2|.$$

This is the sum of the vertical and horizontal distances between x and y (Figure 1.1). You should pause and think about what the ∞-metric on $\mathbb{R}^2$ is measuring.

Many norms, and hence metrics, arise from an "inner product." A (real) *inner product space* $(V, \langle \cdot, \cdot \rangle)$ is a (real) linear space V together with a function $\langle \cdot, \cdot \rangle : V \times V \to \mathbb{R}$ called an *inner product* satisfying five conditions:

(i) $\langle v, v \rangle \geq 0$ for all $v \in V$ (nonnegativity),
(ii) $\langle v, v \rangle = 0$ if and only if $v = 0$ (nondegeneracy),
(iii) $\langle \lambda v, w \rangle = \lambda \langle v, w \rangle$ for all $v, w \in V$ and $\lambda \in \mathbb{R}$ (multiplicativity),
(iv) $\langle v, w \rangle = \langle w, v \rangle$ for all $v, w \in V$ (symmetry),
(v) $\langle v, w + u \rangle = \langle v, w \rangle + \langle v, u \rangle$ for all $u, v, w \in V$ (distributivity).

If $\mathbb{R}$ is replaced by $\mathbb{C}$ everywhere in this definition, and the symmetry property is replaced by the *Hermitian* symmetry property

$$\langle v, w \rangle = \overline{\langle w, v \rangle}$$

for all $v, w \in V$, where the bar indicates the complex conjugate, we get a (complex) inner product space. One must be clear about the underlying field.

Our two basic examples of inner product spaces will not surprise you. Others will be given in the next section.

EXAMPLE 1. The real linear space $\mathbb{R}^n$ with standard inner product

$$\langle x, y \rangle = x_1 y_1 + x_2 y_2 + \cdots + x_n y_n.$$

EXAMPLE 2. The complex linear space $\mathbb{C}^n$ with standard inner product

$$\langle z, w \rangle = z_1 \overline{w_1} + z_2 \overline{w_2} + \cdots + z_n \overline{w_n}.$$

Theorem 1.2. *Inner products always give rise to norms. Specifically, if $(V, \langle \cdot, \cdot \rangle)$ is an inner product space and $\|v\|$ is defined by*

$$\|v\| = \sqrt{\langle v, v \rangle},$$

then $\| \cdot \|$ is a norm on V.

PROOF. Left as Exercise 1.1.6. $\qquad\qquad\qquad\qquad\qquad\qquad\qquad\qquad\square$

Is there an easy way to tell whether a given norm has an inner product associated with it? It turns out that the answer is yes: The *parallelogram equality*,

$$2\|u\|^2 + 2\|v\|^2 = \|u + v\|^2 + \|u - v\|^2,$$

must hold for every pair u and v in an inner product space. On the other hand, if a norm satisfies the parallelogram equality, then it must come from an inner product. Thus, the parallelogram equality characterizes those norms that arise from an inner product. (See Exercise 1.1.8). This equality generalizes the Pythagorean equality and says that the sum of the squares of the lengths of the diagonals of a parallelogram is twice the sum of the squares of the lengths of its sides (Figure 1.2).

Inner products give us a way to talk about a generalized notion of "orthogonality," which will be discussed in detail in Chapter 4. They also give the following very useful theorem.

Theorem 1.3 (Cauchy–Schwarz (or Cauchy–Bunyakovskii–Schwarz) Inequality[2]). *If $(V, \langle \cdot, \cdot \rangle)$ is an inner product space, then*

$$|\langle v, w \rangle| \le \sqrt{\langle v, v \rangle}\sqrt{\langle w, w \rangle}$$

for all $v, w \in V$.

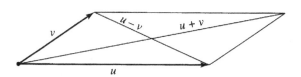

FIGURE 1.2. The parallelogram equality.

---

[2]Named for Viktor Bunyakovskii (1804–1889; Ukraine), Augustin Louis Cauchy (1789–1857; France), and Hermann Schwarz (1843–1921; Poland, now Germany).

PROOF.   We assume that V is a *complex* inner product space. We assume that $w \neq 0$ and first consider the case that $\|w\| = 1$. Then

$$
\begin{aligned}
0 \leq \|v - \langle v, w \rangle w\|^2 &= \langle v - \langle v, w \rangle w, v - \langle v, w \rangle w \rangle \\
&= \langle v, v \rangle - \langle v, w \rangle \overline{\langle v, w \rangle} - \langle v, w \rangle \overline{\langle v, w \rangle} \\
&\quad + \langle v, w \rangle \overline{\langle v, w \rangle} \langle w, w \rangle \\
&= \langle v, v \rangle - \langle v, w \rangle \overline{\langle v, w \rangle} = \|v\|^2 - |\langle v, w \rangle|^2.
\end{aligned}
$$

Therefore, $|\langle v, w \rangle|^2 \leq \|v\|^2$. Now consider an arbitrary $w \neq 0$. Because $w \neq 0$, we have $\|w\| \neq 0$. Therefore, if we let u denote $\frac{w}{\|w\|}$, then $\|u\| = 1$. By the first part of the proof, $|\langle v, u \rangle| \leq \|v\|$. Since $|\langle v, u \rangle| = \frac{|\langle v, w \rangle|}{\|w\|}$, the result is proved. $\square$

This proof works for real inner product spaces as well. However, a somewhat simpler and rather attractive proof can be given for the real case. To see how it goes, we now call the two elements x and y, and so we are trying to prove that

$$
|\langle x, y \rangle| \leq \sqrt{\langle x, x \rangle} \sqrt{\langle y, y \rangle}
$$

in any *real* inner product space. Note that we may assume that $x \neq 0$ and $y \neq 0$. For any real number λ we have

$$
0 \leq \langle \lambda x + y, \lambda x + y \rangle = \lambda^2 \langle x, x \rangle + 2\lambda \langle x, y \rangle + \langle y, y \rangle.
$$

Setting $a = \langle x, x \rangle$, $b = 2\langle x, y \rangle$, and $c = \langle y, y \rangle$ this reads $a\lambda^2 + b\lambda + c \geq 0$ for every $\lambda \in \mathbb{R}$. Since $a > 0$, this quadratic function has a minimum at $\frac{-b}{2a}$, and this minimum value is nonnegative. Thus

$$
a(\frac{-b}{2a})^2 + b(\frac{-b}{2a}) + c \geq 0 \qquad \text{or} \qquad c \geq \frac{b^2}{4a}.
$$

Since $a > 0$, this yields immediately the desired result.

1.2   Examples: Sequence Spaces and Function Spaces

It is the goal of this section to introduce some of the linear spaces that are important to many functional analysts. These will serve as our working examples throughout much of the text.

We first discuss the "sequence spaces." These, as you might guess, are linear spaces whose elements are sequences. The elements can be sequences of real or complex numbers. Addition and scalar multiplication are defined pointwise. There are many sequence spaces; we discuss some of them.

The first example is the collection ℓ^∞ (pronounced "little ell infinity") of all bounded sequences $\{x_n\}_{n=1}^\infty$. The next example is the collection c_0 of all sequences that converge to 0. Notice that $c_0 \subset \ell^\infty$. Both of these collections become normed linear spaces with norm defined by

$$
\|\{x_n\}\|_\infty = \sup\{|x_n| \,|\, 1 \leq n < \infty\}.
$$

(Recall that the *supremum* of a set is its least upper bound, and that the *infimum* of a set is its greatest lower bound.)

Next, we define the ℓ^p-spaces for $1 \leq p < \infty$. The space ℓ^p (pronounced "little ell p") consists of all sequences $\{x_n\}_{n=1}^{\infty}$ such that

$$\sum_{n=1}^{\infty} |x_n|^p < \infty.$$

With norm defined by

$$\|\{x_n\}\|_p = \left(\sum_{n=1}^{\infty} |x_n|^p \right)^{\frac{1}{p}},$$

ℓ^p becomes a normed linear space. Notice that $\ell^p \subset c_0$ for each p, $1 \leq p < \infty$. The space ℓ^1 is somewhat special. It consists of all absolutely convergent sequences. That is, $\{x_n\}_{n=1}^{\infty}$ is in ℓ^1 if and only if the series $\sum_{n=1}^{\infty} |x_n|$ converges. The space ℓ^2 is undoubtedly the most important of all the ℓ^p-spaces for reasons discussed in the last paragraph of this section.

The notation ℓ^p is an abbreviation for $\ell^p(\mathbb{N})$. The reason for this notation is that these spaces are particular examples of the "Lebesgue L^p-spaces." These spaces are named in honor of Henri Lebesgue (1875–1941; France), and you will read much about them in Chapter 3. We only mention this relationship here; you are encouraged to try to understand this relationship after you read Chapter 3.

We now turn to "function spaces." These are linear spaces consisting of functions. As with sequence spaces, addition and scalar multiplication are defined pointwise. The scalars, again, can be taken to be either real or complex. After working with function and sequence spaces for a while you will notice that in some ways the two classes are very much alike. This is perhaps not so surprising if we consider that sequences can, and often should, be thought of as functions defined on $\mathbb{N}$.

Let $[a, b]$ be any closed, bounded interval of $\mathbb{R}$ and let

$$V = \{f : [a, b] \to \mathbb{R} \mid \text{ there exists } B \geq 0 \text{ such that } |f(x)| \leq B \text{ for all } x \in [a, b]\}.$$

This is a linear space. The collection

$$\{f : [a, b] \to \mathbb{R} \mid f \text{ is continuous}\}$$

is a subspace of V. This subspace of all continuous functions on a closed and bounded interval is a very important space in analysis; it is most often denoted by $(C([a, b]), \| \cdot \|_\infty)$, or just $C([a, b])$. With norm defined by

$$\|f\|_\infty = \sup\{|f(x)| \mid x \in [a, b]\},$$

both V and $C([a, b])$ become normed linear spaces.

For each sequence and function space described above we have given a *norm*. According to our results of the previous section, each space is thus a metric space also. On $C([a, b])$, for example, the metric is given by

$$d(f, g) = \|f - g\|_\infty = \sup\{|f(x) - g(x)| \mid x \in [a, b]\}.$$

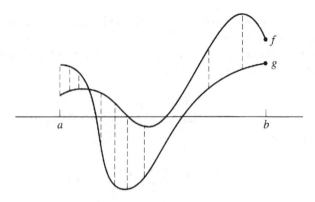

FIGURE 1.3. The supremum metric.

This metric measures the largest vertical distance between the graphs of the two functions (Figure 1.3).

Your next question might be, Which of the norms that we have defined "come from" inner products? We know that a specific norm comes from an inner product if and only if it satisfies the parallelogram equality. Of the norms on sequence spaces discussed above, the ℓ^2-norm is the only one that satisfies the parallelogram equality. Indeed, ℓ^2 is an inner product space with inner product

$$\langle x, y \rangle = \sum_{n=1}^{\infty} x_n y_n$$

if we are considering real-valued functions, and with inner product

$$\langle x, y \rangle = \sum_{n=1}^{\infty} x_n \overline{y_n}$$

if we are considering complex-valued functions. The reader should check that the norm arising from this inner product, via Theorem 1.2, is indeed $\| \cdot \|_2$. If we are considering real-valued functions, the linear space $C([a, b])$ becomes an inner product space with inner product

$$\langle f, g \rangle = \int_a^b f(x)g(x)dx.$$

If we are considering complex-valued functions, $C([a, b])$ becomes an inner product space with inner product

$$\langle f, g \rangle = \int_a^b f(x)\overline{g(x)}dx.$$

Using Theorem 1.2, one can check that the norm associated with this inner product is *not* the supremum norm. The question remains, Does the supremum norm arise from *some* inner product? That is, does it satisfy the parallelogram equality? You will explore the questions raised in this paragraph in the exercises.

1.3   A Discussion About Dimension

Euclidean space $\mathbb{R}^n$ is "finite-dimensional." A "basis" for $\mathbb{R}^n$ is given by the collection of vectors

$$e_1 = (1, 0, 0, \ldots, 0),$$
$$e_2 = (0, 1, 0, \ldots, 0),$$
$$e_3 = (0, 0, 1, \ldots, 0),$$
$$\vdots$$
$$e_n = (0, 0, 0, \ldots, 1).$$

That these vectors form a *basis*[3] for $\mathbb{R}^n$ means

(i) they are *linearly independent*; that is, if $\alpha_1, \alpha_2, \ldots, \alpha_n \in \mathbb{R}$ and

$$\alpha_1 e_1 + \alpha_2 e_2 + \cdots + \alpha_n e_n = 0,$$

then, necessarily, $\alpha_1 = \alpha_2 = \cdots = \alpha_n = 0$;
(ii) they *span* $\mathbb{R}^n$; that is, every vector in $\mathbb{R}^n$ can be written as a linear combination of these basis vectors.

More generally, a linear space is *n-dimensional* if the largest number of linearly independent elements is n. If such an n exists, the space is said to be *finite-dimensional*.

Some of the linear spaces that we have discussed are *infinite-dimensional*. This means that for each positive integer n there exists a linearly independent subset containing n elements. Another way of saying this is that there is no finite subset whose linear combinations span the entire space. In the exercises, you are asked to show that ℓ^1, ℓ^∞, and $C([a, b])$ are infinite-dimensional.

René-Maurice Fréchet was born on September 10, 1878, in Maligny, France (Figure 1.4). He was the fourth of six children. At the time of his birth, his father was the director of a Protestant orphanage. When he was a young boy, his family moved to Paris, where his father became the head of a Protestant school. The French government secularized the schools, and this left Fréchet's father without a job. Eventually, his father again found employment as a teacher, but for a while, his mother took in boarders to bring the family some income. Through the family's boarders, Fréchet became interested in foreign languages and longed to travel. His love of travel lasted for the remainder of his very long life.

Though Fréchet was not financially privileged, he was very fortunate in his schooling. Between the ages of twelve

---

[3]This type of basis is called, more formally, a "Hamel basis." Another notion of basis, an "orthonormal basis" is defined for a Hilbert space. As indicated, this is a different, but related, concept, and we will discuss orthonormal bases in Chapter 4. In the Banach space setting, one can discuss "Schauder bases." We will not go into any detail, but a Schauder basis is defined in the biographical material on Per Enflo.

FIGURE 1.4. Maurice Fréchet.

and fifteen, one of his school teachers was Jacques Hadamard (1865–1963; France). Hadamard was later to become a distinguished mathematician himself, but then he was a young school teacher. Hadamard recognized Fréchet's talents and spent a great deal of extra time with him on mathematics. Hadamard was to have a strong influence over Fréchet's professional life, and the two maintained a close relationship until Hadamard's death.

Fréchet attended École Normale Superieure from 1900 until 1903, after completing military service. He studied both mathematics and physics, eventually choosing to pursue mathematics. The University of Paris awarded him a Ph.D. in 1906.

Fréchet published many papers, the first in 1902, while he was still a stu-

dent. His most significant work is his doctoral dissertation, and this is a true masterpiece.[4] On page 97 of [34] the author asserts that four fundamental papers were written that resulted in the "sudden crystallization of all the ideas and methods which had been slowly accumulating during the nineteenth century." These four papers were Fredholm's 1900 paper on integral equations, Lebesgue's 1902 doctoral dissertation on integration theory, Hilbert's 1906 paper on spectral theory, and Fréchet's 1906 doctoral dissertation on metric spaces.[5] This reference is significant because Fréchet is not thought of as being in the same class as the other three. Though he made other contributions, most notably on linear functionals, best approximation by trigonometric sums, probability, and statistics, it is his thesis work that we will focus on here.

The significance of Fréchet's thesis lies in the fact that it is the first time we see the specific aim of a general theory of metric spaces. The great German mathematician Karl Weierstrass (1815–1897) gave perhaps the first definition of the "nearness" of two functions. This occurred about 1879. Weierstrass's definition was used by the Italian mathematician Vito Volterra (1860–1940). Fréchet, through Hadamard, was influenced by Volterra's work. Fréchet also took as inspiration the work of two other Italians, Giulio Ascoli (1843–1896) and Cesare Arzelà (1874–1912). These two had been working with sets whose elements are functions and were looking to extend the ideas of Georg Cantor

---

[4]In the Introduction we stated that the phrase "functional analysis" was coined by Lévy. Some authors suggest that the true inspiration for this phrase comes from Fréchet's thesis. However, Fréchet himself credits Lévy (see page 260 of [123]).

[5]Erik Ivar Fredholm (1866–1927) was a Swedish mathematician. You will read more about his work in Chapter 5. The others you have already encountered.

44

(1845–1918; Russia) from sets of points to sets consisting of functions.

Fréchet's thesis consists of an introduction and two further parts. Part I contains the rudiments of abstract point set topology. In particular, this includes many results on what Fréchet calls "une class (E)." An E-class was later named a "metric space" (in 1914 by the German mathematician Felix Hausdorff, 1868–1942). Fréchet used the term "écart" for what we now call the "metric," and wrote (x, y) in place of d(x, y). Also in this first part of his thesis, the ideas of "compactness" and "completeness" are formulated. These notions are fundamental to functional analysis, and you will meet them in the next chapter.

In Part II of his thesis Fréchet gives examples to illustrate the theory found in Part I. He uses five examples. The first is Euclidean space $\mathbb{R}^n$, as discussed in the opening paragraph of Section 1 of this chapter. His second example is $C([a, b])$ with écart

$$d(f, g) = \sup\{|f(x) - g(x)| \mid x \in [a, b]\},$$

as we discussed in Section 2. Fréchet's next example is new to us. He lets E_ω denote the set of all sequences and defines the écart on this set by

$$d(x, y) = \sum_{k=1}^{\infty} \frac{1}{k!} \cdot \frac{|x_k - y_k|}{1 + |x_k - y_k|}$$

for two elements $x = \{x_1, x_2, \ldots\}, y = \{y_1, y_2, \ldots\} \in E_\omega$. The fourth example of a metric space consists of a collection of differentiable complex-valued functions, and his last example contains elements that are certain curves in $\mathbb{R}^3$. We will not describe the last two examples.

Fréchet's thesis made a big impact right away. It would be premature to discuss the mathematical details of its impact at this stage of the book. Let us just say that subsequent work of Fréchet in functional analysis was closely related to work of Lebesgue and F. Riesz. In particular, Fréchet did much work on the function space L^2. He was particularly interested in finding necessary and sufficient conditions for a subset of a given metric space to be compact, and he was able to give such conditions for subsets of L^2. You will read much more about the work of Lebesgue and Riesz in later chapters.

After receiving his Ph.D., Fréchet taught first in a high school and then, for short periods, at the Universities of Nantes, Rennes, and Poitiers. From 1914 to 1919 he served in the army, mostly helping with language interpretation between English and French in the battlefields. In 1919 he returned to academic life, in Strasbourg. In 1928 he made his final move, to Paris. He died in Paris, at the age of ninety-four, on June 4, 1973.

Exercises for Chapter 1

Sections 1.1 and 1.2

1.1.1 (a) Verify that $\| \cdot \|_1$, $\| \cdot \|_2$, and $\| \cdot \|_\infty$ define norms on $\mathbb{R}^2$.
(b) To see that these norms are in fact different, compute the distance $d((1, 1), (2, 3))$ (in each of the three norms) between the points $(1, 1)$ and $(2, 3)$.

(c) To see that norms are different, it helps (if it is possible) to visualize the "balls" in the space. Let

$$B_r(x) = \{y \in V \mid \|x - y\| < r\}.$$

This is the *open ball of radius r centered at x*. Of special interest is the "unit ball," $B_1(0) = B_1((0, 0))$. In $\mathbb{R}^2$, with the three different norms, sketch $B_1(0)$ and $B_3((2, 2))$.

1.1.2 Consider the norms $\| \cdot \|_1$ and $\| \cdot \|_\infty$ on $\mathbb{R}^n$.

(a) Prove that

$$\|x\| = \frac{1}{3}\|x\|_1 + \frac{2}{3}\|x\|_\infty$$

defines a norm on $\mathbb{R}^n$.

(b) Sketch the open unit ball in $\mathbb{R}^2$ with respect to this norm.

1.1.3 Verify that ℓ^1 and ℓ^∞ are normed linear spaces.

1.1.4 Verify that $C([a, b])$, with supremum norm, is a normed linear space.

1.1.5 Prove Theorem 1.1.

1.1.6 Prove Theorem 1.2. (Hint: Use the Cauchy–Schwarz inequality to get the triangle inequality.)

1.1.7 Prove that in any complex inner product space

$$\langle v, \lambda w \rangle = \bar{\lambda}\langle v, w \rangle$$

for every $v, w \in V$ and each $\lambda \in \mathbb{C}$.

1.1.8 Consider a normed linear space $(V, \| \cdot \|)$. Recall that some normed linear spaces are inner product spaces, and some are not.

(a) Prove that the parallelogram equality characterizes the inner product spaces among the normed linear spaces. That is, show that

$$\|v\| = \sqrt{\langle v, v \rangle}$$

for some inner product $\langle \cdot, \cdot \rangle$ if and only if

$$2\|u\|^2 + 2\|v\|^2 = \|u + v\|^2 + \|u - v\|^2$$

holds for every pair u and v in V.

(b) Is the parallelogram equality satisfied in ℓ^1? In ℓ^∞?

1.1.9 Use the preceding exercise to show that the supremum norm on $C([a, b])$ cannot come from an inner product.

1.1.10 (a) In $C([0, 1])$ with supremum norm, compute $d(f, g)$ for $f(x) = 1$ and $g(x) = x$.

(b) Repeat part (a), with the supremum norm replaced by the norm induced by the inner product

$$\langle f, g \rangle = \int_0^1 f(x)g(x)dx.$$

Section 1.3

1.3.1 The basis for $\mathbb{R}^n$ given in the text is called the "standard basis." It is a basis, but not the only basis, for $\mathbb{R}^n$. Give another basis for $\mathbb{R}^3$, and show that it is in fact a basis.

1.3.2 Prove that ℓ^1 is infinite-dimensional. Explain why your proof also shows that ℓ^∞ is infinite-dimensional.

1.3.3 Prove that $C([0, 1])$ is infinite-dimensional.

2

The Topology of Metric Spaces

2.1 Open, Closed, and Compact Sets; the Heine–Borel and Ascoli–Arzelà Theorems

Let (M, d) be a metric space. Recall that the *r-ball centered at* x is the set

$$B_r(x) = \{y \in M \,|\, d(x, y) < r\}$$

for any choice of $x \in M$ and $r > 0$. These sets are most often called *open balls*, *open disks*, or *open neighborhoods*, and they are denoted by the above or by $B(x, r)$, $D_r(x)$, $D(x, r)$, $N_r(x)$, $N(x, r)$, among other notations. A point $x \in M$ is a *limit point* of a set $E \subseteq M$ if every open ball $B_r(x)$ contains a point $y \neq x$, $y \in E$. If $x \in E$ and x is not a limit point of E, then x is an *isolated point* of E. E is *closed* if every limit point of E is in E. A point x is an *interior point* of E if there exists an $r > 0$ such that $B_r(x) \subseteq E$. E is *open* if every point of E is an interior point. A collection of sets is called a *cover* of E if E is contained in the union of the sets in the collection. If each set in a cover of E is open, the cover is called an *open cover* of E. If the union of the sets in a subcollection of the cover still contains E, the subcollection is referred to as a *subcover* for E. E is *compact* if every open cover of E contains a finite subcover. E is *sequentially compact* if every sequence of E contains a convergent subsequence. E is *dense* in M if every point of M is a limit point of E. The *closure* of E, denoted by $\overline{E}$, is E together with its limit points. The *interior* of E, denoted by E° or int(E), is the set of interior points of E. E is *bounded* if for each $x \in E$, there exists $r > 0$ such that $E \subseteq B_r(x)$.

You should watch for differences in the literature concerning these definitions. Some look different but really aren't, but some are actually different. In particular, be careful with the limit point definition. For example, Hoffman and Marsden ([89] page 145) allow that y can equal x. Thus, their limit point of E can be in E. It can be an isolated point of E. They use *accumulation point* to refer to our limit point.

We begin with three basic topological results. Their proofs are left as exercises.

Theorem 2.1. *A set E in a metric space (M, d) is open if and only if its complement, $E^c = M \setminus E$, is closed.*

Theorem 2.2.

(a) *For any collection $\{E_\alpha\}$ of open sets, $\bigcup_\alpha E_\alpha$ is open.*
(b) *For any collection $\{F_\alpha\}$ of closed sets, $\bigcap_\alpha F_\alpha$ is closed.*
(c) *For any finite collection $\{E_i\}_{i=1}^n$ of open sets, $\bigcap_{i=1}^n E_i$ is open.*
(d) *For any finite collection $\{F_i\}_{i=1}^n$ of closed sets, $\bigcup_{i=1}^n F_i$ is closed.*

Theorem 2.3. *If E is a compact subset of a metric space, then E is closed.*

Of the types of sets defined above, compact sets are particularly interesting. The idea of a general notion of compactness is to get at the quality possessed by a closed and bounded interval of $\mathbb{R}$ that forces every continuous function to attain its maximum and minimum on the interval. The notion of compactness is usually hard for students to use at first. Showing that a given set is *not* compact can be straightforward (see Exercise 2.1.10). On the other hand, using the definition to show that a given set *is* compact can be quite tricky, since to do so, one must consider *every* open cover of the set. Thus, we are interested in characterizing the compact subsets of our favorite metric spaces. The easiest characterization is known as the Heine–Borel theorem (Theorem 2.5) which describes the compact subsets of $\mathbb{R}^n$. You may be familiar with the proof, in at least the case $n = 1$. The Heine–Borel theorem gives a very nice characterization of the compact subsets of $\mathbb{R}^n$. Are there any other metric spaces in which the compact subsets can be characterized so nicely? The Ascoli–Arzelà theorem (Theorem 2.6) describes the compact subsets of $C([a, b])$. Recall from Fréchet's biography that he gave another such characterization (specifically, of the compact subsets of the metric space L^2, which will be defined in Chapter 3). The proofs of the Heine–Borel theorem and the Ascoli–Arzelà theorem both use the next result, which gives an equivalent, and very useful, condition for compactness.

Theorem 2.4. *A subset of a metric space is compact if and only if it is sequentially compact.*

PROOF. Let M be a metric space and $E \subseteq M$ be compact. Assume that there exists a sequence $\{x_n\}_{n=1}^\infty$ of E with no convergent subsequence. Then, among the x_n's, there are infinitely many distinct points. Call them $\{y_n\}_{n=1}^\infty$. Let U_k be an open set containing y_k yet containing no other y_n. Since the set $\{y_n\}_{n=1}^\infty$ has no limit points, it is closed, and hence $M \setminus \{y_n\}_{n=1}^\infty$ is open. Then the U_k's together with $M \setminus \{y_n\}_{n=1}^\infty$ form an open cover of E (in fact of M). Because E is compact, this cover has a finite subcover, say

$$U_1, \ldots, U_N, M \setminus \{y_n\}_{n=1}^{\infty}.$$

Then $U_1, \ldots, U_N$ is a finite subcover of the set $\{y_n\}_{n=1}^{\infty}$. This contradicts the construction of the U_k's, and thus we conclude that in fact, E is sequentially compact.

To show the other implication we assume that E is sequentially compact and that $\{U_\alpha\}$ is an arbitrary open cover of E. We aim to show that $\{U_\alpha\}$ contains a finite subcover.

First, suppose that for each positive integer n we can choose a $y_n \in E$ such that $B_{\frac{1}{n}}(y_n)$ is not contained in any U_α. By hypothesis, $\{y_n\}_{n=1}^{\infty}$ has a convergent subsequence, say $z_n \to z \in E$. Note that $z \in U_{\alpha_0}$ for some U_{α_0}. Choose $\epsilon > 0$ such that $B_\epsilon(z) \subseteq U_{\alpha_0}$. Choose N large enough so that $d(z_n, z) < \frac{\epsilon}{2}$ for $n > N$, and $\frac{1}{N} < \frac{\epsilon}{2}$. Then $B_{\frac{1}{N}}(z) \subseteq U_{\alpha_0}$, a contradiction. Thus, there exists $r > 0$ such that for every $y \in E$, $B_r(y) \subseteq U_\alpha$ for some U_α.

Second, suppose that there exists $\epsilon > 0$ such that E cannot be covered by finitely many ϵ-balls. Construct a sequence $\{y_n\}_{n=1}^{\infty}$ inductively as follows: Let y_1 be any element of E; choose $y_2 \in E \setminus B_\epsilon(y_1)$, $y_3 \in E \setminus [B_\epsilon(y_1) \cup B_\epsilon(y_2)]$, and so on. Then $d(y_n, y_m) > \epsilon$ for all n and m. Thus $\{y_n\}_{n=1}^{\infty}$ has no convergent subsequence. This contradicts that E is sequentially compact. Thus, for each $\epsilon > 0$ we can cover E with finitely many ϵ-balls.

Finally, let $r > 0$ be as above. We know that we can cover E with finitely many r-balls. Let $x_1, \ldots, x_n$ denote their centers. Then each $B_r(x_k)$ is contained in a U_{α_k} for some U_{α_k}. The collection $U_{\alpha_1}, \ldots, U_{\alpha_n}$ is the desired finite subcover of the U_α's. This completes the proof.   □

In the above proof we proved that if $A \subseteq B \subseteq M$, where M is any metric space, A is closed, and B is compact, then A is compact. We point this out because this result is useful on its own.

If a set has the property that for each $\epsilon > 0$ we can cover E with a finite number of ϵ-balls, then the set is said to be *totally bounded*. This property appeared in the proof of Theorem 2.4, and it was shown that any compact (equivalently, sequentially compact) set is totally bounded. It is straightforward to show that a totally bounded set is always bounded but that the converse is not so (see Exercise 2.1.12).

The next theorem is named in honor of Emile Borel (1871–1956; France) and Eduard Heine (1821–1881; Germany).

Theorem 2.5 (The Heine–Borel Theorem).   *A subset E of $\mathbb{R}^n$ is compact if and only if it is closed and bounded.*

PROOF.   Let $E \subseteq \mathbb{R}^n$ be compact. E is closed by Theorem 2.3, and E is bounded by the remarks in the paragraph preceding the statement of this theorem.

To prove the converse, we assume that E is closed and bounded, show that E is sequentially compact, and apply Theorem 2.4. Let $\{x_k\}_{k=1}^{\infty}$ be a sequence in E.

Each x_k is an n-tuple of real numbers, and we can display them in an array:

$$
\begin{array}{ccccc}
x_1^1 & x_1^2 & x_1^3 & \cdots & x_1^n \\
x_2^1 & x_2^2 & x_2^3 & \cdots & x_2^n \\
x_3^1 & x_3^2 & x_3^3 & \cdots & x_3^n \\
\vdots & \vdots & \vdots & \vdots & \vdots
\end{array}
$$

Since E is bounded, every number appearing in this array is smaller than some fixed real number. Thus, each column is a bounded sequence of real numbers, and hence contains a convergent subsequence. In particular, the first column, $\{x_k^1\}_{k=1}^\infty$ contains a convergent subsequence. Denote it by $\{x_{k_j}^1\}_{j=1}^\infty$. Then the corresponding subsequence $\{x_{k_j}^2\}_{j=1}^\infty$ of the second column contains a convergent subsequence. Continue, producing finally a convergent subsequence of the last column. To simplify notation, denote this subsequence by $\{x_{k_j}^n\}_{j=1}^\infty$. Then $\{x_{k_j}^l\}_{j=1}^\infty$ converges for each $l = 1, \ldots, n$. Therefore, $\{x_{k_j}\}_{j=1}^\infty$ is a convergent subsequence of $\{x_k\}_{k=1}^\infty$. Since E is assumed closed, its limit point is in E. This completes the proof that E is sequentially compact, and hence the proof of the theorem. □

Let $E \subseteq C([a,b])$. The set E is *equicontinuous at* $x \in [a,b]$ if for any $\epsilon > 0$ there exists a $\delta > 0$ such that $y \in [a,b]$ and $|x - y| < \delta$ imply $|f(x) - f(y)| < \epsilon$ for all $f \in E$. The set E is *equicontinuous* if it is equicontinuous at each $x \in [a,b]$.

In this definition, $C([a,b])$ denotes all *real-valued* continuous functions defined on the compact interval $[a,b]$. Let us denote this, temporarily, by $C([a,b]; \mathbb{R})$. We can replace $[a,b]$ by any subset of any arbitrary metric space and $\mathbb{R}$ by any normed linear space. That is, if (M, d) is a metric space, $A \subseteq M$, and $(V, \|\cdot\|)$ is a normed space, then $C(A; V)$ is a linear space, and definitions such as "equicontinuous" can be extended to this setting. Also, theorems, like the Ascoli–Arzelà theorem can be generalized with due caution (for example, the Ascoli–Arzelà theorem requires A to be compact, as $[a,b]$ was). If the set V needs to be emphasized, we will continue to write $C(A; V)$. Otherwise, we will write this collection of functions in abbreviated form $C(A)$. You may wonder why we cannot replace the normed space V by simply a metric space; we cannot do this because we need to be able to *add* the elements of V.

Notice that the definition of "equicontinuous" is similar to that of a set of functions being uniformly continuous: δ is chosen independent of x_0 (as it was in uniform continuity) and now it is also independent of f.

We are now ready to give a characterization of the compact subsets of $C([a,b])$. This theorem is referred to as the Ascoli–Arzelà theorem, the Arzelà–Ascoli theorem, or Ascoli's theorem. Giulio Ascoli (1843–1896; Italy) published a proof in 1883. Cesare Arzelà (1874–1912; Italy) re-proved the result independently, and published his proof in 1895, giving credit to Ascoli. There are many generalizations of this theorem, and these constitute a class of results known as "Ascoli theorems." You might also, now, wonder about the naming of the Heine–Borel theorem. The discovery of this theorem was gradual, and more complicated. We mention that Heine's proof was given in 1872, the year after Borel was born, and recommend that you see, for example, page 953 of [76] for a more complete rendering of the story of the theorem's development.

Theorem 2.6 (The Ascoli–Arzelà Theorem). *Let $E \subseteq C([a, b]; \mathbb{R})$. Then E is compact if and only if E is closed, bounded, and equicontinuous.*

PROOF.   Assume that E is closed, bounded, and equicontinuous. We aim to show that E is compact. It suffices to show, by Theorem 2.4, that E is sequentially compact. To this end, let $\{f_n\}$ be a sequence in E. We aim to find a convergent subsequence.

For any $\delta > 0$, the collection of δ-balls $\{B_\delta(x) : a \le x \le b\}$ covers $[a, b]$, and since this interval is compact (by the Heine–Borel theorem), there exists a finite subcover, say

$$B_\delta(x_1), \ldots, B_\delta(x_n).$$

Let $X_\delta = \{x_1, \ldots, x_n\}$ and $X = \bigcup_{k=1}^{\infty} X_{\frac{1}{k}}$. Since each $X_{\frac{1}{k}}$ is finite, X is countable. Let $X = \{y_1, y_2, \ldots\}$ be an enumeration of X.

Consider the sequence of real numbers $\{f_n(y_1)\}_{n=1}^{\infty}$. Because E is bounded, there exists a number M such that $|f(x)| < M$ for all $f \in E$ and all $x \in [a, b]$. In particular, the sequence $\{f_n(y_1)\}_{n=1}^{\infty}$ is in the interval $[-M, M]$. Since this interval is compact, it is sequentially compact, and hence $\{f_n(y_1)\}$ has a convergent subsequence. Denote this subsequence by

$$f_{11}(y_1), f_{12}(y_1), f_{13}(y_1), \ldots.$$

Similarly, $\{f_{1n}(y_2)\}_{n=1}^{\infty}$ has a convergent subsequence, which we will denote by

$$f_{21}(y_2), f_{22}(y_2), f_{23}(y_2), \ldots.$$

Continuing, the sequence $\{f_{2n}(y_3)\}_{n=1}^{\infty}$ has a convergent subsequence, which we will denote by

$$f_{31}(y_3), f_{32}(y_3), f_{33}(y_3), \ldots.$$

Continue this process. Set $g_n = f_{nn}$. We have

$$
\begin{array}{cccc}
f_{11} & f_{12} & f_{13} & \cdots \\
f_{21} & f_{22} & f_{23} & \cdots \\
f_{31} & f_{32} & f_{33} & \cdots \\
\cdots & \cdots & \cdots & \cdots \\
\cdots & \cdots & \cdots & \cdots
\end{array}
$$

where the sequence given in any particular row is a subsequence of the row right above it. By construction,

$$\lim_{n \to \infty} f_{kn}(y_m)$$

exists for each specified value of k and m. Thus, the sequence of functions $\{g_n\}$ converges at each point of X.

The proof of this direction will be complete when we show that $\{g_n\}_{n=1}^{\infty}$ converges at each point of $[a, b]$ and that this convergence is uniform. To do this, let $\epsilon > 0$ and let δ be as in the definition of equicontinuity. Let X_δ be as before. Since each of the sequences $\{g_k(x_1)\}_{k=1}^{\infty}, \{g_k(x_2)\}_{k=1}^{\infty}, \dots \{g_k(x_n)\}_{k=1}^{\infty}$ converges,) **why?** there exist N_j's such that

$$|g_n(x_j) - g_m(x_j)| < \frac{\epsilon}{3}$$

whenever $n, m > N_j$. Let $N = \max\{N_1, \dots, N_n\}$. Let $x \in [a, b]$ be arbitrary and $x_j \in X_\delta$ be such that $|x - x_j| < \delta$. By the definition of equicontinuity,

$$|g_n(x) - g_n(x_j)| < \frac{\epsilon}{3}$$

for all n. Therefore,

$$|g_n(x) - g_m(x)| \leq |g_n(x) - g_n(x_j)| + |g_n(x_j) - g_m(x_j)| + |g_m(x_j) - g_m(x)|$$
$$< \frac{\epsilon}{3} + \frac{\epsilon}{3} + \frac{\epsilon}{3} = \epsilon$$

whenever $n, m > N$ and $x \in [a, b]$. This shows that $d(g_n, g_m) < \epsilon$ for all $n, m > N$ and hence that the sequence $\{g_n\}_{n=1}^{\infty}$ is uniformly convergent. Since E was assumed closed, the limit of this sequence also lies in E. This proves that E is sequentially compact.

For the other direction, assume that E is compact. Then E is closed (by Theorem 2.3) and bounded, and in fact, E is totally bounded (by the remarks preceding the Heine–Borel theorem). It remains to be shown that E is equicontinuous. Let $\epsilon > 0$. Because E is totally bounded, we can cover E with $\frac{\epsilon}{3}$-balls $B_{\frac{\epsilon}{3}}(f_1), \cdots, B_{\frac{\epsilon}{3}}(f_n)$. For $x \in [a, b]$ choose $\delta > 0$ such that $|f_i(x) - f_i(y)| < \epsilon$ holds for all $y \in B_\delta(x) = (x - \delta, x + \delta)$ and all $i = 1, \dots, n$. Let $y \in B_\delta(x)$ and $f \in E$ be arbitrary. Then there exists an i with $f \in B_{\frac{\epsilon}{3}}(f_i)$ and

$$|f(x) - f(y)| \leq |f(x) - f_i(x)| + |f_i(x) - f_i(y)| + |f_i(y) - f(y)|$$
$$< \frac{\epsilon}{3} + \frac{\epsilon}{3} + \frac{\epsilon}{3} = \epsilon.$$

This shows that E is equicontinuous at x; since x was arbitrary in $[a, b]$, the proof is complete. $\square$

We end this section by proving that the closed unit ball of $C([0, 1])$ is closed and bounded, but is not compact. We do this by showing that the closed unit ball is not equicontinuous. It is also possible to prove that it is not compact by showing that it is not sequentially compact. This alternative proof is left as an exercise (Exercise 2.1.13(c)). Let D denote the closed unit ball of $C([0, 1])$. It is clear that D is bounded. If f_0 is a limit point of D, then for each positive integer n there exists and $f_n \in D$ with $d(f, f_n) < \frac{1}{n}$. Then

$$\|f_0\| = d(f_0, 0) \leq d(f_0, f_n) + d(f_n, 0) < \frac{1}{n} + 1.$$

Since this inequality holds for all n, $\|f_0\| \leq 1$. This shows that D is closed.

Finally, consider $\{f_n\}_{n=1}^{\infty} \subseteq D$, where $f_n(x) = x^n$ for each n. If D were compact, then D would be equicontinuous and hence so would be this subset of D. Let $\epsilon = \frac{1}{2}$. By the definition of equicontinuity there exists a $\delta > 0$ such that $|x - y| < \delta$ and $x, y \in [0, 1]$ imply that $d(x^n, y^n) < \frac{1}{2}$ for every n. In particular, $|1 - y^n| < \frac{1}{2}$ for all n whenever $|1 - y| < \delta$. Since $y \in [0, 1]$, $y^n \to 0$ as $n \to \infty$. If we choose n large enough so that $y^n < \frac{1}{2}$, we do not have $|1 - y^n| < \frac{1}{2}$. We have thus shown that D cannot be compact.

In Exercise 2.1.13 you will explore the compactness of unit balls for a few other spaces. In fact, the compactness of the closed unit ball characterizes finite-dimensional normed linear spaces. This result was proved by Frigyes Riesz (1880–1956; Austria-Hungary, now Hungary).[1]

2.2   Separability

A metric space is *separable* if it contains a countable dense subset.

EXAMPLE 1. $\mathbb{R}$, endowed with the Euclidean metric, is separable because $\mathbb{Q}$ is a countable set and is dense in $\mathbb{R}$. Likewise, $\mathbb{R}^n$ is separable because $\mathbb{Q}^n$ is a countable set and is dense in $\mathbb{R}^n$.

EXAMPLE 2. $C([a, b])$ is separable. This follows from the Weierstrass approximation theorem (Theorem 6.1), which states that

$$\mathcal{P}([a, b]) = \{f \in C([a, b]) \mid f \text{ is a polynomial with real coefficients}\}$$

is dense in $C([a, b])$. This result is not trivial. Next, let

$$\mathcal{Q}([a, b]) = \{f \in C([a, b]) \mid f \text{ is a polynomial with rational coefficients}\}.$$

One can show that

  (i) $\mathcal{Q}([a, b])$ is countable, and
  (ii) $\mathcal{Q}([a, b])$ is dense in $\mathcal{P}([a, b])$.

Observations (i) and (ii) together imply that $C([a, b])$ is separable.
In practical terms, the fact that $\mathbb{Q}$ dense in $\mathbb{R}$ means that, for example, we can approximate π to any desired degree of accuracy. That $\mathcal{Q}([a, b])$ is dense in $C([a, b])$ means that we can always use a polynomial with rational coefficients to approximate, as closely as desired, a given continuous function on $[a, b]$.

EXAMPLE 3. $\mathbb{R}$ with the discrete metric is not separable. To see this, suppose that $\{x_k\}_{k=1}^{\infty}$ is a countable dense subset and that $x \in \mathbb{R}$ is not an element of this sequence. Then $d(x, x_k) = 1$ for each k, and so $B_{\frac{1}{2}}(x)$ cannot contain an element of the sequence. This contradicts the hypothesis that the sequence is dense.

---

[1] Frigyes Riesz's brother Marcel was also a distinguished mathematician.

2.3   Completeness: Banach and Hilbert Spaces

In Exercise 1.1.9 you were asked to prove that the supremum norm on $C([a, b])$ does not come from an inner product. Recall, however, that we can endow this linear space with an inner product via

$$\langle f, g \rangle = \int_a^b f(x)g(x)dx.$$

The induced norm is then

$$\|f\|_2 = \sqrt{\int_a^b |f(x)|^2 dx},$$

and not

$$\|f\|_\infty = \sup\{|f(x)| \,\big|\, x \in [a, b]\}.$$

This might suggest that the supremum norm is, in some sense, less desirable than the norm $\| \cdot \|_2$ on $C([a, b])$. There are advantages and disadvantages to each norm. The supremum norm does have the very nice property that it is "complete" (Theorem 2.7) on $C([a, b])$, while the norm $\| \cdot \|_2$ on $C([a, b])$ is not complete.

Recall that a sequence $\{x_n\}_{n=1}^\infty$ of real numbers is said to be Cauchy if given any $\epsilon > 0$ there exists an integer $N > 0$ such that $|x_n - x_m| < \epsilon$ whenever $n, m \geq N$. More generally, a sequence $\{x_n\}_{n=1}^\infty$ in a metric space is *Cauchy* if given any $\epsilon > 0$ there exists an integer $N > 0$ such that $d(x_n, x_m) < \epsilon$ whenever $n, m \geq N$.

It is easily seen that any convergent sequence is Cauchy: Assume that $\{x_n\}_{n=1}^\infty$ converges and let $\epsilon > 0$. There is thus an $x \in M$ and a positive integer N such that $d(x_n, x) < \frac{\epsilon}{2}$ for all $n \geq N$. Then

$$d(x_n, x_m) \leq d(x_n, x) + d(x, x_m) < \frac{\epsilon}{2} + \frac{\epsilon}{2} = \epsilon$$

for all $n, m \geq N$.

You may remember that in $\mathbb{R}$ the converse of this statement is also true. There are metric spaces, however, in which this converse does not hold (see Exercise 2.3.6). A subset A of a metric space is called *complete* if every Cauchy sequence in A converges to a point of A. If a metric space is complete in itself, the space is called a *complete metric space*. A complete normed linear space is called a *Banach space*. A complete inner product space is called a *Hilbert space*.

It is not clear who coined the phrase "Banach space." Banach himself referred to these spaces as "espaces-B." However, shortly after Banach published his monumental book [11] in 1932, the terminology became standard. Banach's book was a comprehensive account of all known results at the time on normed linear spaces. John von Neumann is credited with originating, in the late 1920s, the current usage of "Hilbert space." However, perhaps as early as 1904–1905 Hilbert's students called ℓ^2 "Hilbert's space" (see [117]).

Theorem 2.7. $C([a, b]; \mathbb{R})$ *with the supremum norm is complete.*

PROOF.   Let $\{f_n\}_{n=1}^{\infty}$ be a Cauchy sequence in $C([a, b]; \mathbb{R})$. This means that given any $\epsilon > 0$ there exists an integer $N > 0$ such that $\|f_n - f_m\|_{\infty} < \frac{\epsilon}{2}$ whenever $n, m \geq N$. That is, given any $\epsilon > 0$ there exists an integer $N > 0$ such that $|f_n(x) - f_m(x)| < \frac{\epsilon}{2}$ for all $n, m \geq N$ and each $x \in [a, b]$. Then $\{f_k(x)\}_{k=1}^{\infty}$ is a Cauchy sequence of real numbers for each $x \in [a, b]$. Thus $\{f_k(x)\}_{k=1}^{\infty}$ converges to some real number for each x; we will denote this value by $f(x)$. This defines a new function f such that $f_n \to f$ pointwise. It remains to be shown that $\{f_n\}_{n=1}^{\infty}$ converges uniformly to this f, and that f is continuous. Since the uniform limit of a sequence of continuous functions is again continuous, it remains to show that $\{f_n\}_{n=1}^{\infty}$ converges uniformly to f. Let $\epsilon > 0$. Then there exists an N such that $|f_m(x) - f_n(x)| < \frac{\epsilon}{2}$ for every choice of $x \in [a, b]$ and $n, m > N$. If $m \geq N$ and $x \in [a, b]$, then

$$f_n(x) \in \left(f_m(x) - \frac{\epsilon}{2}, f_m(x) + \frac{\epsilon}{2}\right), \text{ for all } n \geq N.$$

Therefore,

$$f(x) \in \left[f_m(x) - \frac{\epsilon}{2}, f_m(x) + \frac{\epsilon}{2}\right],$$

and hence

$$|f(x) - f_m(x)| \leq \frac{\epsilon}{2} < \epsilon.$$

Since $x \in [a, b]$ was arbitrary, we are done.   □

It turns out that $[a, b]$ in the statement of Theorem 2.7 can be replaced by any *compact* subset of any topological space, and $\mathbb{R}$ can be replaced by any *complete* metric space.

Two of the names most commonly associated with the development of function and sequence spaces are Stefan Banach and David Hilbert. It would be difficult to overestimate the influence that their work has had in the field.

David Hilbert (Figures 2.1 and 2.2) was born on January 23, 1862, in Königsberg, Prussia (now Kaliningrad, Russia). He was from a family that kept good records, and there is much known about his ancestry. The book [104] is highly recommended. Hilbert's father was a judge at the time of Hilbert's birth, and his mother is considered to have had a strong intellectual leaning. David Hilbert was the first of two children. Hilbert did not start school until the age of eight (two years late by standards of the time). It is not clear why he did not, but it is probable that he was at first "home schooled" by his mother. His school career was good, but there seems to be nothing outstanding about it.

In 1880 Hilbert enrolled at the University of Königsberg. He was at once a devoted and hardworking student of mathematics. During his university years he studied mostly at Königsberg, but also at Heidelberg, and received his Ph.D. from Königsberg in 1885. Hilbert's entire life was marked by stability. This stability existed in his family life, and in his professional life. He remained in Königsberg, working at the University until

FIGURE 2.2. Hilbert in 1937.

FIGURE 2.1. A 1912 portrait of Hilbert, sold as part of a group of postcards of Göttingen professors.

1895, when he moved to the University of Göttingen. He spent the rest of his life in Göttingen.

Hilbert approached mathematics in a somewhat unusual way. He would completely devote himself to one area for a rather intense period, and then turn, again with intensity, to another area. Often, his choice of topics seemed to be in very different areas of mathematics. He also spent a period working in physics. He was a very "public" worker, always sharing ideas with others and often working on collaborative efforts. He had sixty-nine doctoral students.

The first area that Hilbert worked in was invariant theory. He pursued this area during, roughly, the period 1885–1893. In fact, he proved the most important open problem at the time: Gordan's problem. His solution contained a very early example of an existence, as opposed to a constructive, proof. At the time, such a proof was viewed with suspicion and not accepted by many mathematicians. Ultimately, Hilbert also provided a constructive proof to Gordan's problem.

In 1893 he turned his attention to number theory. He worked in this area for about five years. During this time he gave proofs of the transcendence of π and e; both had earlier been proved transcendental, but Hilbert's proofs offered considerable improvements. In 1897 his *Zahlbericht* was published. The goal of this approximately 400 page book was to summarize the current state of number theory. Not only did it achieve that goal, but it contained ideas that would lead to the development of new mathematics, including the entire not-yet-born field of homological algebra. After laying a solid foundation that others would subsequently build on, Hilbert turned his attention away from number theory.

During this same time he became interested in the so-called Dirichlet principle. This principle gave, loosely, a method for solving certain boundary value problems. It had become well known when Riemann used it repeatedly in his 1851 doctoral dissertation. The trouble was that this principle had not been proved to work in all cases. Weierstrass objected to

Riemann's use of this principle because he suspected that it was not always applicable. Weierstrass was correctly skeptical, and he was able to prove that it does not always work. However, in 1899 Hilbert showed that it does work if certain conditions are satisfied. Hilbert's results salvaged Riemann's work but, unfortunately, only after Riemann's death.

Every four years mathematicians meet at the International Congress of Mathematicians. This is a huge gathering, and since 1936 the Fields Medals have been awarded at it. Hilbert was invited to give an address at the second International Congress of Mathematicians, held in Paris in the summer of 1900. The speech that he gave is perhaps the most famous mathematical speech ever given.[2] Hilbert had a few ideas of what to speak on. In consultation with his good friend Hermann Minkowski (1864–1901; Russia)[3] he decided to speak on the future of mathematics. The planned speech consisted of a list of 23 problems on which mathematicians should focus their energies over the next century. As it turned out, he had planned too much to say and ended up, after some introductory remarks, discussing 10 of his problems. A statement of the problems, and updates of progress on their solutions, can be found in various articles and texts (see, for example, [52]) and web sites (including through links found at the web site listed in the Introduction). These problems are known by their numbers. At the 1900 Congress, Hilbert discussed problems 1, 2, 6, 7, 8, 13, 16, 19, 21, and 22. The first three of these are about the foundations of mathematics, the next four about algebraic topics, and the last three about function theory. The list includes famous problems such as the continuum hypothesis (1) and Riemann's hypothesis about the location of the zeros of the zeta function (8). The first problem to be solved was the third one, and its solution appeared already in 1900. Others have been solved in full, some partially, and others remain unsolved. You will read a bit about Hilbert's fifth problem in Chapter 5, when we discuss the contemporary mathematician Per Enflo (born 1944; Sweden).

The 23 problems covered a broad range of the mathematical topics that were current at the time. However, there are omissions. In particular, it seems odd that there are really no questions having to do with what was soon to be called "functional analysis." At the time of Hilbert's speech, ideas of functional analysis had been floating around for about a decade. Very soon after the Congress, Hilbert himself was to become deeply involved in the birth of the field.

It is the years 1902–1912 that are of most interest to us, for these are the years that Hilbert devoted to integral equations. It is out of this work that functional analysis was born. His interest in this subject was sparked by a paper by Fredholm on integral equations. In this paper, Hilbert recognized a link being made between integral equations and what we now call linear algebra. Fredholm's

---

[2]This speech has certainly had an impact on twentieth-century mathematics. An argument can be made that in addition to its positive impact, it had some negative influence as well. For a discussion of some of the negative effect it has had, see the recent article [52].

[3]The friendship and collaboration between Hilbert and Minkowski is quite interesting, but we do not have the space to go into it here.

efforts were related to work on oscillating systems, work that originated with Fourier. We will not discuss Fourier's work at all here; you will read much more about it later in the text. Simply put, one can take a problem of finding solutions to an integral equation and rephrase it as a problem of finding "eigenfunctions" for a given "linear operator." These "eigenfunctions" are elements of some Hilbert space. By the end of this course you should understand very clearly what is meant by these last sentences. Let us just say that realizing this connection was brilliant, and it led to the opening up of an entire field of mathematics that has been more fruitful than could possibly have been imagined at its birth.

From 1910 to 1922 Hilbert worked on problems of physics. In particular, he was working on field equations for a general theory of relativity. Though his ideas foresaw some of the famous ideas that were to come later, to others, the work of Albert Einstein (1879–1955; Germany) ultimately proved more successful than Hilbert's. See [104] for more on this work, and the Einstein–Hilbert priority dispute.

Around the end of this period the new "quantum theory" was also being developed. Though Hilbert did not directly work in this field, his earlier work on integral equations, equations involving infinitely many variables, and eigenfunctions turned out to be very useful for this new area. In Hilbert's own words [104], "I developed my theory of infinitely many variables from purely mathematical interests, and even called it 'spectral analysis' without any presentiment that it would later find an application to the actual spectrum of physics."

During the periods that we have omitted, Hilbert worked in geometry and on the foundations of mathematics. He made significant contributions to these fields, but we will not discuss them.

During Hilbert's time there, Göttingen was a world center for mathematics and mathematical physics. Carl Gauss, Peter Dirichlet, and Bernhard Riemann had all worked there. When Hilbert arrived, much activity was centered around Felix Klein (1849–1925; Prussia, now Germany). Klein drew students from all over the world, particularly from the United States. The University at Göttingen continued to flourish, and it is amazing to see the list of people who spent substantial amounts of time there during Hilbert's tenure; this list includes, but is not limited to, Otto Blumenthal, Harald Bohr, Max Born, Constantin Carathéodory, Richard Courant, Werner Heisenberg, Ernst Hellinger, Edmund Landau, Hermann Minkowski, Emmy Noether, Carl Runge, Erhard Schmidt, Otto Toeplitz, Hermann Weyl, and Ernst Zermelo.

The dispersal of this group is a sad story. In 1930 Hilbert took mandatory retirement. During the time that the Göttingen center was forming, it was difficult for certain individuals to get jobs. This included Jews and women. Hilbert was very open about these matters, and held no prejudices. He gathered around him the people that he wanted to work with independent of their race, sex, or religious beliefs, in some cases, fighting vigorously to be allowed to invite an individual to Göttingen. As a result, the University of Göttingen was an obvious target for "cleansing" by the Nazis. By 1933, Göttingen was essentially emptied of its mathematicians and physicists. Many of these individuals fled to the United States, and it is a story of great historical irony that American science benefited so much from the Nazi program.

David Hilbert died in Göttingen on February 14, 1943.

Stefan Banach was born thirty years after Hilbert, on March 30, 1892, in Cracow, Poland (Figure 2.3). In contrast to Hilbert, there is not much known about Banach's parents, nor about his earliest years. It is known that his father was a civil servant, and that his mother gave him up on April 3, 1892. Apparently, his early childhood was spent with his paternal grandmother.

Banach lacked a "traditional" academic education, attended university irregularly, eventually passing his half-diploma exams (equivalent to first and second year university) in 1914 at the Lvov Technical University. Without the privilege afforded by a more traditional education, Banach did not proceed to an academic post in the usual fashion. Instead, he was "discovered" by another famous Polish mathematician, Hugo Steinhaus. The story of Steinhaus's discovery of Banach is amusing, and is told in [73] and [79]. Banach's talents were then quickly recognized by the larger Polish mathematical community, and he shortly thereafter completed his Ph.D. at the Jan Kazimierz University. In 1922 he became a professor at the University of Lvov. Due to the strengths of Banach and Steinhaus, Lvov became a very important international center for functional analysis at that time.

Banach was prolific over the next decade, and he became the world's leading authority on functional analysis. Several important theorems bear his name. We will study Banach's contraction mapping principle and the Banach–Steinhaus and Hahn–Banach theorems (all in Chapter 6). Most importantly, he systematized the theory of functional analysis. His efforts neatly pulled together isolated results due, primarily, to Fredholm, Hilbert, and Volterra on integral equations. Banach's

FIGURE 2.3. Stefan Banach.

most important and lasting contributions are his monumental book *Théorie des Opérations Linéaires*, published in 1932, and his founding of the journal *Studia Mathematica* in 1929. This internationally respected journal was started by Banach and Steinhaus and was, and is, devoted to publishing articles having to do with functional analysis. In his book, many of the notions of modern functional analysis were introduced, including the axioms for normed linear spaces and the idea of a dual space.

Although functional analysis was Banach's main area, he also made contributions to other fields. Indeed, the famous Banach-Tarski paradox is a startling result about set theory [12]. Banach was also very active in the "Scottish Café" group; I recommend reading more about this café, and the mathematics that went on there.

Although he was born decades after Hilbert, the end of Banach's life was also marked by the terror of the Nazi regime. Banach's story is more tragic. During

World War II, Lvov was first occupied by the Soviets, and then by the Germans. The Soviets deported most of the Poles from Lvov, but Banach was very much repected by them, and he managed to stay in Lvov. When the Germans took control of the city in 1941 matters changed dramatically for Banach. Most professors were arrested and sent to concentration camps. Though Banach survived the Nazi order to eliminate the elite, he suffered greatly during the coming years. He spent time during the German occupation as a lice feeder in an institute producing vaccines against typhoid. When the Soviets returned to Lvov in 1944, Banach returned to mathematics, accepting a chair at the Jagiellonian University in Cracow (one of Europe's oldest universities, founded in 1364, and alma mater to Copernicus). However, Banach's health was by now poor, and he died on August 31, 1945.

Exercises for Chapter 2

Section 2.1

2.1.1 Prove Theorem 2.1.

2.1.2 Prove Theorem 2.2.

2.1.3 Give examples to show that the word *finite* cannot be omitted in the last two parts of Theorem 2.2.

2.1.4 Prove Theorem 2.3. Remember that your proof should be in a general metric space.

2.1.5 (a) Consider $\mathbb{R}$ with the Euclidean metric. Is $(0, 1)$ open, closed, neither, or both? Explain. (It may surprise you to learn that a set can be neither open nor closed, or it can be both (a *clopen* set). In the rest of this exercise set you will meet examples of such sets.)

(b) Consider $\mathbb{R}^2$ with the Euclidean metric. Is $(0, 1) = \{(x, 0) \mid 0 < x < 1\}$ open, closed, neither, or both? Explain.

2.1.6 Is it true that $\text{int}(A \cup B) = \text{int}(A) \cup \text{int}(B)$? Prove or give a counterexample. What if union is replaced by intersection?

2.1.7 Let M be any set with the discrete metric. What are the open sets?

2.1.8 Consider $\mathbb{R}$ with the Euclidean metric.

(a) Find the set of limit points of $\{\frac{1}{n} \mid n = 1, 2, 3, \ldots\}$.

(b) Find the set of limit points of $\mathbb{Q}$.

2.1.9 Give an example of a closed and bounded set in a metric space that is not compact.

2.1.10 Consider $\mathbb{R}$ with the Euclidean metric. Give an open cover of $(-10, 10]$ that has no finite subcover.

2.1.11 This exercise is about the famous set due to Georg Cantor. If you do not know a definition of the Cantor set, look it up in a real analysis text or ask your professor for a reference.

(a) Is the Cantor set compact? Explain.

(b) What is the interior of the Cantor set? Explain.

2.1.12 Prove that a totally bounded set is necessarily bounded. Then, using the discrete metric on any set of your choice, show that a bounded set need not be totally bounded.

2.1.13 (a) Show that the closed unit ball in $C([0, 1])$ is not compact by showing that it is not sequentially compact. That is, construct a sequence of functions with norm less than or equal to one that does not have a convergent subsequence.

(b) Show that the closed unit ball in ℓ^1 (respectively ℓ^∞) is not compact.

(c) Show that the closed unit ball in any infinite-dimensional space is not compact.

2.1.14 Show that if M is a metric space, $A \subseteq M$ is compact, and $U \subseteq M$ is open, then $A \setminus U$ is compact.

2.1.15 Consider metric spaces M_1 and M_2.

(a) Prove that a function $f : M_1 \to M_2$ is continuous if and only if the inverse image of every open set in M_2 is open in M_1.

(b) Assume that $A \subseteq M$ is compact and that $f : M \to \mathbb{R}$ is continuous. Show that $f(A)$ is a bounded subset of $\mathbb{R}$.

Section 2.2

2.2.1 Complete the proof outlined in Example 2 to show that $C([a, b])$ is separable.

2.2.2 Is ℓ^1 separable? Prove your assertion.

2.2.3 Is ℓ^∞ separable? Prove your assertion.

Section 2.3

2.3.1 Prove that in a metric space every Cauchy sequence is bounded.

2.3.2 Let $\| \cdot \|$ and $\vertiii{ \cdot }$ denote two norms defined on the same linear space X. Suppose that there exist constants $a > 0$ and $b > 0$ such that

$$a\|x\| \le \vertiii{x} \le b\|x\| \tag{2.1}$$

for each $x \in X$. Show that $(X, \| \cdot \|)$ is complete if and only if $(X, \vertiii{ \cdot })$ is complete. If two norms satisfy (2.1) they are called *equivalent norms*.

2.3.3 (a) On $\mathbb{R}^n$, show that the three norms $\| \cdot \|_1$, $\| \cdot \|_2$, and $\| \cdot \|_\infty$ are equivalent.

(b) Prove that in a finite-dimensional linear space all norms are equivalent. You will notice that (a) follows from (b); still, please do (a) first. It is instructive.

2.3.4 Let M be a set with the discrete metric. Is M complete? Explain.

2.3.5 Define

$$\|f\|_1 = \int_a^b |f(x)| dx$$

on $C([a, b])$. Show that this does, indeed, define a norm. Does this norm come from an inner product? Is $(C([a, b]), \| \cdot \|_1)$ a Banach space?

2.3.6 Show that $(C([a, b]), \| \cdot \|_2)$ is not complete.

2.3.7 Prove that ℓ^1 is complete.

2.3.8 Prove that ℓ^∞ is complete.

3
Measure and Integration

The foundations of integration theory date to the classical Greek period. The most notable contribution from that time is the "method of exhaustion" due to Eudoxos (ca. 408–355 B.C.E.; Asia Minor, now Turkey). Over two thousand years later, Augustin Cauchy stressed the importance of defining an integral as a limit of sums. One's first encounter with a theory of the integral is usually with a variation on Cauchy's definition given by Georg Friedrich Bernhard Riemann (1826–1866; Hanover, now Germany). Though the Riemann integral is attractive for many reasons and is an appropriate integral to learn first, it does have deficiencies. For one, the class of Riemann integrable functions is too small for many purposes. Henri Lebesgue gave, around 1900, another approach to integration. In addition to the integrals of Riemann and Lebesgue, there are yet other integrals, and debate is alive about which one is the best. Arguably, there is no one best integral. Different integrals work for different types of problems. It can be said, however, that Lebesgue's ideas have been extremely successful, and that the Lebesgue integrable functions are the "right" ones for many functional analysts and probabilists. It is no coincidence that the rapid development of functional analysis coincided with the emergence of Lebesgue's work.

As you will discover in this chapter, Lebesgue's ideas on integration are intertwined with his notion of measure. It was this idea — of using measure theory as a platform for integration — that marked a departure from what had been done previously. Lebesgue's measure theory and application to integration theory appear in his doctoral thesis [80]. This thesis is considered one of the greatest mathematical achievements of the twentieth century.

The sole reason that this material on measure and integration is included in this book is because of the important role it plays in functional analysis. Frigyes Riesz was familiar with Lebesgue's work, and also with David Hilbert's work on integral equations and Maurice Fréchet's work on abstract function spaces. He combined these elements brilliantly, developing theories now considered basic to the field of functional analysis. Riesz deserves much credit for recognizing the importance of Lebesgue's ideas and drawing attention to them. Riesz's work will be introduced in the last section of this chapter, and further explored in the next chapter, culminating with the celebrated Riesz–Fischer theorem.

We start our study of measure by considering some problems from probability. Probability theory had, in some sense, been in the mathematical world since the mid-seventeenth century. However, it wasn't until the 1930s, when Andrei Kolmogorov (1903–1987; Russia) laid the foundation for the theory using Lebesgue's measure theory, that probability was truly viewed as a branch of pure mathematics. It is therefore historically inaccurate to use probability to motivate measure theory. Nonetheless, the applications of measure theory to probability theory are beautiful, and they provide very good source of inspiration for students about to embark on their first journey into the rather technical field of measure theory.

3.1   Probability Theory as Motivation

In this section we give an informal introduction to Lebesgue's theory of measure, using an example from probability as inspiration. The ideas for the presentation of this material come from [1]. Consider a sequence of coin tosses of a fair coin. We represent such a sequence by, for example,

$$\text{THHTTTHTHTT}\ldots. \tag{3.1}$$

Let

$$s_n = \text{the number of heads in } n \text{ tosses.}$$

The law of large numbers asserts, in some sense or other, that the ratio $\frac{s_n}{n}$ approaches $\frac{1}{2}$ as n gets larger. The goal of this section is to rephrase this law in measure-theoretic language, thus indicating that measure theory provides a "framework" for probability. Although measure theory did not arise because of probability theorists "looking for a language," one could argue that probability theory helped to ensure measure theory's importance as a branch of pure mathematics worthy of research efforts. We will let this discussion of very basic probability theory serve as a source of motivation for learning about measure theory, and not for continuing the discussion about probability theory. A further investigation of probability theory is a worthy endeavor, but it would take us too far afield to do it here.

The law of large numbers was first stated in the seventeenth century by James (=Jakob=Jacques) Bernoulli (1654–1705; Switzerland). In his honor a sequence, like (3.1) of independent trials with two possible outcomes is called a *Bernoulli sequence*. Let $\mathcal{B}$ denote the collection of all Bernoulli sequences. This is the so-called sample space from probability theory, and is often denoted by Ω in that context.

We note that $\mathcal{B}$ is uncountable. This can be shown using a Cantor diagonalization argument. An alternative, and more useful, proof proceeds by showing that all but a countable number of the elements of $\mathcal{B}$ can be put in to one-to-one correspondence with the interval $(0, 1]$. Specifically, let $\mathcal{B}_T$ denote the set of Bernoulli trials that are constantly T after a while. Then $\mathcal{B}_T$ is countable (left as an exercise), and so $\mathcal{B} \setminus \mathcal{B}_T$ is still uncountable if $\mathcal{B}$ is. The reader is asked to write out the details of both proofs that $\mathcal{B}$ is uncountable in Exercises 3.1.1 and 3.1.2. To a Bernoulli sequence associate the real number ω whose binary expansion is $\omega = .a_1 a_2 a_3 \dots$, where $a_i = 1$ if the ith toss in the sequence is a head, and $a_i = 0$ if the ith toss in the sequence is a tail. In the case that ω has two binary expansions, exactly one will be nonterminating, and we use this one. Thus,

$$\omega = \sum_{i=1}^{\infty} \frac{a_i}{2^i}.$$

This allows us to identify subsets E of $\mathcal{B}$ ("events") with subsets B_E of $(0, 1]$.

In the general theory of probability, the sample space Ω can be any set. If Ω is finite or countably infinite, Ω is called a *discrete probability space*, and the probability theory is relatively straightforward. However, if Ω is uncountably infinite, as $\mathcal{B}$ is, then the sets B_E can be quite complicated. It is for determining the "size" of these sets B_E that we require measure theory.

Heuristically, a *measure* μ on a space Ω should be a nonnegative function defined on certain subsets of Ω (hereafter referred to as *measurable sets*). For a measurable set A, $\mu(A)$ denotes the measure of A. As a guiding principle of sorts, we require that

$$\mu \left(\bigcup_{i=1}^{n} A_i \right) = \sum_{i=1}^{n} \mu(A_i)$$

for every finite collection $\{A_i\}_{i=1}^{n}$ of measurable sets satisfying $A_i \cap A_j = \emptyset$ $(i \neq j)$.

We will begin our study of abstract measures with a specific example of a measure: *Lebesgue measure*, often denoted by μ_L or m, on certain subsets of $\Omega = \mathbb{R}$. We will require that $m(I) = b - a$ if I is any one of the four intervals (a, b), $(a, b]$, $[a, b)$, and $[a, b]$. It follows from this that each finite subset of $\mathbb{R}$ has Lebesgue measure zero. Every countable set will be seen to have Lebesgue measure zero, and we will see that there are even uncountably infinite sets with Lebesgue measure zero (the Cantor set is such a set, see Exercise 3.2.10).

We now discuss the connection between probability and measure theory. We associate subsets E of $\mathcal{B}$ with subsets B_E of the interval $(0, 1]$, as before. We then *define* the *probability that the event E occurs*, Prob(E), to be the Lebesgue measure $m(B_E)$ of the set B_E. Using this, let us look at two very basic examples and see that the value of $m(B_E)$ agrees with what we would expect Prob(E) to be from everyday experience.

EXAMPLE 1.   Let E be the event that a head is thrown on the first toss. We know that Prob(E) should equal $\frac{1}{2}$. Let's now figure out what the set B_E is, and then see whether this set has Lebesgue measure $\frac{1}{2}$. A number ω is in B_E if and only if $\omega = 0.1a_2a_3\dots$. Therefore, ω is in B_E if and only if $\omega \geq 0.1000\dots$ and $\omega \leq 0.1111\dots$. That is, $B_E = [\frac{1}{2}, 1]$. Then $m(B_E) = \frac{1}{2}$, as desired.

EXAMPLE 2.   In the first example we considered the event that the first toss is prescribed. This time we let E be the event that the first n tosses are prescribed. Let us say that these first n tosses are a_1, a_2, a_3, ..., a_n. We know that Prob(E) should equal $\left(\frac{1}{2}\right)^n$. As in the first example, we now try to identify the set B_E, and figure out its Lebesgue measure. If we let $s = 0.a_1a_2a_3\dots a_n00000\dots$, then ω is in B_E if and only $\omega \geq s$ and $\omega \leq 0.a_1a_2a_3\dots a_n11111\dots$. But

$$0.a_1a_2a_3\dots a_n11111\dots = s + \sum_{i=n+1}^{\infty} \frac{a_i}{2^i} = s + \left(\frac{1}{2}\right)^n.$$

Therefore, $B_E = \left[s, s + \left(\frac{1}{2}\right)^n\right]$, and $m(B_E) = \left(\frac{1}{2}\right)^n$, as desired.

We now return to the law of large numbers. We will give two versions of this result. Since the material of this section is primarily offered to motivate a study of measure, proofs are not included. First, for $\omega \in (0, 1]$, we define $s_n(\omega) = a_1 + \cdots + a_n$, where $0.a_1a_2\dots$ is the binary representation for ω.

Theorem (Weak Law of Large Numbers).   *Fix $\epsilon > 0$ and define, for each positive integer n, the set*

$$B_n = \left\{\omega \in (0, 1] \,\middle|\, \left|\frac{s_n(\omega)}{n} - \frac{1}{2}\right| > \epsilon\right\}.$$

This subset of $(0, 1]$ corresponds to the event that "after the first n tosses, the number of heads is not close to $\frac{1}{2}$." The weak law of large numbers states that $m(B_n) \to 0$ as $n \to 0$.

Theorem (Strong Law of Large Numbers).   *Let*

$$S = \left\{\omega \in (0, 1] \,\middle|\, \lim_{n\to\infty} \frac{s_n(\omega)}{n} = \frac{1}{2}\right\}.$$

The strong law of large numbers states that $m((0, 1] \setminus S) = 0$.

We end this section by remarking that the set $(0, 1] \setminus S$ in the strong law of large numbers is uncountable. After the Cantor set, this is our second example of an uncountable set of Lebesgue measure zero.

3.2   Lebesgue Measure on Euclidean Space

Before we give a formal treatment of Lebesgue measure on $\mathbb{R}^n$, we give a few general definitions. A family $\mathcal{R}$ of sets is called a *ring* if $A \in \mathcal{R}$ and $B \in \mathcal{R}$ imply $A \cup B \in \mathcal{R}$ and $A \setminus B \in \mathcal{R}$ (we remark that B need not be a subset of A in order to define $A \setminus B$). A ring $\mathcal{R}$ is called a σ-*ring* if $A_k \in \mathcal{R}$, $k = 1, 2, \dots$, implies

$\bigcup_{k=1}^{\infty} A_k \in \mathcal{R}$. We consider functions μ defined on a ring or σ-ring $\mathcal{R}$ and taking values in $\mathbb{R} \cup \{\pm\infty\}$. Such a function μ is *additive* if

$$\mu(A \cup B) = \mu(A) + \mu(B)$$

whenever $A \cap B = \emptyset$. If for each sequence $A_k \in \mathcal{R}, k = 1, 2, \ldots$, with $\bigcup_{k=1}^{\infty} A_k \in \mathcal{R}$ we have

$$\mu\left(\bigcup_{k=1}^{\infty} A_k\right) = \sum_{k=1}^{\infty} \mu(A_k)$$

whenever $A_k \cap A_j = \emptyset$ ($k \neq j$), we say that μ is *countably additive*. We must assume that the range of μ does not contain both ∞ and $-\infty$, or else the right side of $\mu(A \cup B) = \mu(A) + \mu(B)$ might not make sense. A countably additive, nonnegative function μ defined on a ring $\mathcal{R}$ is called a *measure*. The elements of $\mathcal{R}$ are subsets of some set X; X is called the *measure space*.

In general, measures can (and do) exist on rings consisting of subsets of any set. We will see examples of such abstract measures in the last section of this chapter. Until then we restrict ourselves to Euclidean space $\mathbb{R}^n$, and develop Lebesgue measure on a (yet-to-be-specified) ring of subsets of $\mathbb{R}^n$.

We consider subsets of $\mathbb{R}^n$ of form

$$\{(x_1, \ldots, x_n) \mid a_k \leq x_k \leq b_k, k = 1, \ldots, n\},$$

where $(a_1, \ldots, a_n)$ and $(b_1, \ldots, b_n)$ are fixed elements in $\mathbb{R}^n$ with each $a_k \leq b_k$. Often we use the notation

$$[a_1, b_1] \times \cdots \times [a_n, b_n]$$

to denote the set just described. Any or all of the $\leq$ signs may be replaced by $<$, with corresponding changes made in the interval notation. Such subsets of $\mathbb{R}^n$ are called the *intervals* of $\mathbb{R}^n$. For an interval I, we define the Lebesgue measure $m(I)$ of I by

$$m(I) = \prod_{k=1}^{n} (b_k - a_k).$$

This definition is independent of whether $\leq$s or $<$s appear in the definition of I. It should be noted that if $n = 1, 2,$ or 3, then m is the length, area, or volume of I. We can extend m to $\mathcal{E}$, the collection of all finite unions of disjoint intervals, by requiring that m be additive. In Exercise 3.2.4 you are asked to show that $\mathcal{E}$ is a ring. Note that $m(A) < \infty$ for any $A \in \mathcal{E}$.

Lemma 3.1. *If $A \in \mathcal{E}$ and $\epsilon > 0$, then there exists a closed set $F \in \mathcal{E}$ and an open set $G \in \mathcal{E}$ such that $F \subseteq A \subseteq G$ and*

$$m(F) \geq m(A) - \epsilon \quad \text{and} \quad m(G) \leq m(A) + \epsilon.$$

PROOF.   Left as Exercise 3.2.5.   □

Theorem 3.2. *m is a measure on $\mathcal{E}$.*

PROOF.   All that needs to be shown is that m is countably additive. Suppose that $\{A_k\}_{k=1}^{\infty}$ is a disjoint collection of sets in $\mathcal{E}$ and that $A = \bigcup_{k=1}^{\infty} A_k$ is also in $\mathcal{E}$. For each N, $\bigcup_{k=1}^{N} A_k \subseteq A$, and so (by Exercise 3.2.2(a))

$$m(A) \geq m\left(\bigcup_{k=1}^{N} A_k\right) = \sum_{k=1}^{N} m(A_k).$$

Since this holds for each N,

$$m(A) \geq \sum_{k=1}^{\infty} m(A_k).$$

We now aim to show that the other inequality also holds. Choose a closed set F corresponding to A as in Lemma 3.1 to satisfy

$$m(F) \geq m(A) - \epsilon.$$

Choose an open set G_k for each A_k as in Lemma 3.1 satisfying

$$m(G_k) \leq m(A_k) + \frac{\epsilon}{2^k}.$$

Notice that F is closed and bounded, and thus is compact by the Heine–Borel theorem. Since $\{G_k\}_{k=1}^{\infty}$ is an open cover for F, there exists an integer N such that

$$F \subseteq G_1 \cup G_2 \cup \cdots \cup G_N.$$

Then

$$m(A) - \epsilon \leq m(F) \leq m(G_1 \cup \cdots \cup G_N) \leq m(G_1) + \cdots + m(G_N)$$
$$\leq m(A_1) + \cdots + m(A_N) + \epsilon.$$

From this,

$$m(A) \leq \sum_{k=1}^{N} m(A_k) + 2\epsilon \leq \sum_{nk=1}^{\infty} m(A_k) + 2\epsilon.$$

Since $\epsilon > 0$ was arbitrary,

$$m(A) \leq \sum_{k=1}^{\infty} m(A_k),$$

completing the proof.   □

We have now constructed Lebesgue measure m on $\mathcal{E}$ consisting of certain subsets of $\mathbb{R}^n$. The reader should verify that $\mathcal{E}$ is a ring, but not a σ-ring (Exercise 3.2.4). We would like to extend m to a much larger (σ-)ring of subsets of $\mathbb{R}^n$. First, we point out that for any set X, the collection of all subsets of X is a ring (and is even a σ-ring). This ring is often denoted by 2^X. To extend m to a larger ring than $\mathcal{E}$ we proceed by first extending m to an "outer measure" m^* defined on all of the ring $2^{(\mathbb{R}^n)}$. Unfortunately, m^* will not actually be a measure on $2^{(\mathbb{R}^n)}$ (hence the new name "outer measure"). We will then take a certain collection $\mathcal{M}$ such that

$\mathcal{E} \subseteq \mathcal{M} \subseteq 2^{(\mathbb{R}^n)}$. Happily, $\mathcal{M}$ will be big enough to be a σ-ring, and small enough so that m^* restricted to $\mathcal{M}$ will be a measure.

Let A be any subset of $\mathbb{R}^n$ and consider a countable covering of A with intervals I_k such that $A \subseteq \bigcup_{k=1}^{\infty} I_k$. We define the *outer measure $m^*(A)$* of A by

$$m^*(A) = \inf \sum_{k=1}^{\infty} m(I_k),$$

where the infimum is taken over all such coverings of A. We call m^* the outer measure corresponding to m.

Note that m^* is defined on all of $2^{(\mathbb{R}^n)}$. It should be clear to the reader that m^* is nonnegative and monotone (that is, $A \subseteq B$ implies $m^*(A) \le m^*(B)$), and that $m(A) = m^*(A) < \infty$ for $A \in \mathcal{E}$. Also, the reader should check (and is given the opportunity in Exercise 3.2.6) that m^* is *countably subadditive*, that is,

$$m^*\left(\bigcup_{k=1}^{\infty} A_k\right) \le \sum_{k=1}^{\infty} m^*(A_k),$$

whenever $A_1, A_2, \ldots$ are subsets of $\mathbb{R}^n$.

For two sets A, B in $\mathbb{R}^n$, we define their *symmetric difference*

$$S(A, B) = (A \setminus B) \cup (B \setminus A)$$

and the distance from A to B by

$$D(A, B) = m^*\Big(S(A, B)\Big).$$

We let $\mathcal{M}_{\mathcal{F}}$ denote the collection of subsets A of $\mathbb{R}^n$ such that $D(A_k, A) \to 0$ as $k \to \infty$ for some sequence of sets $A_k \in \mathcal{E}$. We let $\mathcal{M}$ denote the collection of subsets of $\mathbb{R}^n$ that can be written as a countable union of sets in $\mathcal{M}_{\mathcal{F}}$. It should be evident that $\mathcal{M}_{\mathcal{F}} \subseteq \mathcal{M}$.

As a precursor to the next lemma, we point out that m^* satisfies a sort of continuity condition. Consider two subsets A, B of $\mathbb{R}^n$ with at least one of $m^*(A)$ and $m^*(B)$ finite; we assume that $m^*(B) < \infty$ and that $m^*(B) < m^*(A)$. Then

$$m^*(A) = D(A, \emptyset) \le D(B, \emptyset) + D(A, B) = m^*(B) + D(A, B).$$

Therefore,

$$|m^*(A) - m^*(B)| \le D(A, B).$$

Lemma 3.3. *m^* is additive on $\mathcal{M}_{\mathcal{F}}$.*

PROOF. Let A and B be disjoint sets in $\mathcal{M}_{\mathcal{F}}$. We aim to prove that

$$m^*(A \cup B) = m^*(A) + m^*(B).$$

We choose sets A_k, B_k in $\mathcal{E}$ such that $D(A_k, A) \to 0$ and $D(B_k, B) \to 0$ as $k \to \infty$. From Exercise 3.2.7(e) we have

$$D(A_k \cup B_k, A \cup B) \to 0 \qquad \text{as } n \to \infty.$$

Since m^* restricted to $\mathcal{E}$ coincides with m, we can make two observations (see Theorem 3.2):

$$m^*(A_k \cup B_k) = m^*(A_k) + m^*(B_k)$$

for each k, and all three terms in this equation are finite. Using these two observations and the continuity property of m^* we have

$$|m^*(A \cup B) - m^*(A) - m^*(B)|$$
$$\leq |m^*(A \cup B) - m^*(A_k \cup B_k)| + |m^*(A_k) - m^*(A)| + |m^*(B_k) - m^*(B)|$$
$$\leq D(A_k \cup B_k, A \cup B) + D(A_k, A) + D(B_k, B).$$

Since all three terms of this sum tend to zero as $k \to \infty$, we are done. □

Lemma 3.4. $\mathcal{M}_{\mathcal{F}}$ *is a ring.*

PROOF.   Consider A, B in $\mathcal{M}_{\mathcal{F}}$ and A_k, B_k in $\mathcal{E}$ such that $D(A_k, A) \to 0$ and $D(B_k, B) \to 0$ as $k \to \infty$. Then, for each k, $A_k \cup B_k \in \mathcal{E}$ by Exercise 3.2.4, and by Exercise 3.2.7(e),

$$D(A_k \cup B_k, A \cup B) \to 0 \qquad \text{as } k \to \infty,$$

showing that $A \cup B$ is in $\mathcal{M}_{\mathcal{F}}$. It remains to be seen that $A \setminus B$ is $\mathcal{M}_{\mathcal{F}}$. We now do this; this proof should give an idea of how to do Exercise 3.2.7(d). Again, since $\mathcal{E}$ is a ring, $A_k \setminus B_k$ is in $\mathcal{E}$. From

$$S(A_k \setminus B_k, A \setminus B) = S(A_k \cap B_k^c, A \cap B^c)$$
$$\subseteq S(A_k, A) \cup S(B_k^c, B^c)$$
$$= S(A_k, A) \cup S(B_k, B),$$

it follows that

$$D(A_k \setminus B_k, A \setminus B) = m^*\left(S(A_k \setminus B_k, A \setminus B)\right)$$
$$\leq m^*\left(S(A_k, A)\right) + m^*\left(S(B_k, B)\right)$$
$$= D(A_k, A) + D(B_k, B).$$

This proves that $A \setminus B$ is $\mathcal{M}_{\mathcal{F}}$, as desired. □

Lemma 3.5. *Let $A \in \mathcal{M}$. Then $A \in \mathcal{M}_{\mathcal{F}}$ if and only if $m^*(A) < \infty$.*

PROOF.   First, assume that $A \in \mathcal{M}_{\mathcal{F}}$. Then there exists a sequence of sets $A_k \in \mathcal{E}$ satisfying $D(A_k, A) \to 0$ as $k \to \infty$. We choose N to satisfy $D(A_N, A) < 1$. From Exercise 3.2.7(d) it follows that

$$D(A, \emptyset) \leq D(A, A_N) + D(A_N, \emptyset),$$

or

$$m^*(A) \leq D(A, A_N) + m^*(A_N) < 1 + m^*(A_N) < \infty.$$

To prove the converse we assume that $A \in \mathcal{M}$ and that $m^*(A) < \infty$. We aim to show that $A \in \mathcal{M}_{\mathcal{F}}$. Since $A \in \mathcal{M}$, we can write $A = \bigcup_{k=1}^{\infty} B_k$, where $B_k \in \mathcal{M}_{\mathcal{F}}$

for each k. Letting $A_1 = B_1$, and $A_k = B_k \setminus \left(\bigcup_{j=1}^{k-1} B_j \right)$ for $k \geq 2$, we note that $A_k \in \mathcal{M}_{\mathcal{F}}$ for each k, and we have rewritten A as the union $A = \bigcup_{k=1}^{\infty} A_k$ of disjoint sets $A_1, A_2, \ldots$.

Countable subadditivity of m^* (see Exercise 3.2.6) yields

$$m^*(A) \leq \sum_{k=1}^{\infty} m^*(A_k).$$

We claim that this is actually an equality. To see this, note that for each N,

$$\bigcup_{k=1}^{N} A_k \subseteq A,$$

and so, by monotonicity,

$$m^*\left(\bigcup_{k=1}^{N} A_k \right) \leq m^*(A).$$

Lemma 3.3 asserts that m^* is additive when restricted to $\mathcal{M}_{\mathcal{F}}$. Therefore,

$$\sum_{k=1}^{N} m^*(A_k) = m^*\left(\bigcup_{k=1}^{N} A_k \right).$$

Since now

$$\sum_{k=1}^{N} m^*(A_k) \leq m^*(A)$$

for each N, we have shown that

$$\sum_{k=1}^{\infty} m^*(A_k) \leq m^*(A).$$

We are assuming that $m^*(A) < \infty$, and hence the series on the left converges. Therefore, given $\epsilon > 0$, there exists an N such that $\sum_{k=N+1}^{\infty} m^*(A_k) < \epsilon$. Then

$$D\left(A, \bigcup_{k=1}^{N} A_k \right) = m^*\left(S\left(A, \bigcup_{k=1}^{N} A_k \right) \right)$$

$$= m^*\left(A \setminus \bigcup_{k=1}^{N} A_k \right)$$

$$= m^*\left(\bigcup_{k=N+1}^{\infty} A_k \right) \leq \sum_{k=N+1}^{\infty} m^*(A_k) < \epsilon.$$

Since ϵ was chosen arbitrarily, this proves that $A \in \mathcal{M}_{\mathcal{F}}$.   □

Theorem 3.6. $\mathcal{M}$ *is a σ-ring, and m^* is countably additive on $\mathcal{M}$.*

PROOF.   First, we prove that $\mathcal{M}$ is a σ-ring.

If $A_1, A_2, \ldots \in \mathcal{M}$, then their union can be seen to be in $\mathcal{M}$ via a standard diagonalization argument.

Let $A, B \in \mathcal{M}$. We can then write

$$A = \bigcup_{k=1}^{\infty} A_k, \quad \text{and} \quad B = \bigcup_{k=1}^{\infty} B_k,$$

for some $A_k, B_k \in \mathcal{M_F}, k = 1, 2, \ldots$. The reader should check that the identity

$$A_k \cap B = \bigcup_{j=1}^{\infty} (A_k \cap B_j), \quad k = 1, 2 \ldots,$$

holds. From this it follows that $A_k \cap B \in \mathcal{M}$ for each k. Since

$$m^*(A_k \cap B) \leq m^*(A_k) < \infty,$$

Lemma 3.5 implies that $A_k \cap B \in \mathcal{M_F}$ for each n. Lemma 3.4 then implies that

$$A_k \setminus B = A_k \setminus (A_k \cap B) \in \mathcal{M_F}, \quad k = 1, 2 \ldots.$$

Finally we have our desired result, that

$$A \setminus B = \bigcup_{k=1}^{\infty} \left(A_k \setminus B \right) \in \mathcal{M}.$$

Second, we prove that m^* is countably additive on $\mathcal{M}$. We consider $A = \bigcup_{k=1}^{\infty} B_k$, where $B_k \in \mathcal{M}$ for each k (and so also $A \in \mathcal{M}$ by the first part of the theorem). Letting $A_1 = B_1$, and $A_k = B_k \setminus \left(\bigcup_{j=1}^{k-1} B_j \right)$ for $k \geq 2$, we can rewrite A as the union $A = \bigcup_{k=1}^{\infty} A_k$ of disjoint sets $A_1, A_2, \ldots$. In Exercise 3.2.6 you are asked to prove that m^* is countably subadditive, and therefore

$$m^*(A) \leq \sum_{k=1}^{\infty} m^*(A_k).$$

On the other hand, for each positive integer N,

$$\bigcup_{k=1}^{N} A_k \subseteq A,$$

so the additivity of m^* on $\mathcal{M_F}$ (Lemma 3.3) and the monotonicity of m^* on $2^{(\mathbb{R}^n)}$ together imply that

$$\sum_{k=1}^{N} m^*(A_k) \leq m^*(A).$$

Since this holds for each positive integer N,

$$\sum_{k=1}^{\infty} m^*(A_k) = m^*(A),$$

as desired. □

We now have rings $\mathcal{E} \subseteq \mathcal{M} \subseteq 2^{(\mathbb{R}^n)}$ and an outer measure m^* defined on $2^{(\mathbb{R}^n)}$. Further, m^* restricted to the σ-ring $\mathcal{M}$ is a measure. The elements of $\mathcal{M}$ are called the *Lebesgue measurable subsets of* $\mathbb{R}^n$. The restriction of m^* to $\mathcal{M}$ is called *Lebesgue measure,* and is (again) denoted by m. It is important to figure out which subsets of $\mathbb{R}^n$ are Lebesgue measurable. Exercises 3.2.9 and 3.2.10 give some answers. After doing that exercise you may well wonder whether there are any sets in $\mathbb{R}^n$ that are not Lebesgue measurable, and if there are, just how bizarre they must be. There are indeed such sets. A discussion of nonmeasurable sets is deferred to the Section 6.4. Because the Lebesgue measurable sets are hard to describe, people often choose to work with Lebesgue measure on the smaller σ-ring of all "Borel" sets. The Borel sets are defined, and briefly discussed, in the second example of Section 6 of this chapter.

FIGURE 3.1. Henri Lebesgue.

Henri Léon Lebesgue was born on June 28, 1875, in Beauvais, France (Figure 3.1). His father was a typographical worker, and his mother was an elementary-school teacher; both were intellectually motivated people. In 1897 Lebesgue graduated from the École Normale Supérieure in Paris and then worked for two years in their library. During these two years, he published his first four mathematical papers. His first paper gave a simpler proof of the Weierstrass approximation theorem (discussed in detail in Section 6.1).

Lebesgue can be said to have made two huge contributions to mathematics: He helped to sort out the correct definition of the term "function," and he developed complete, and to date the most successful, theories of measure and integration. For historical perspective, [93] is a good reference for the former contribution, and [61] is recommended for the latter. In his obituary of Lebesgue [29], J.C. Burkill concludes, "His work lay almost entirely in one field — the theory of real functions; in that field he is supreme."

Between 1899 and 1902 Lebesgue was teaching at the lycée in Nancy, and also working on his thesis. During these three years he published six papers. The last five of these were then incorporated to form his doctoral thesis. He received his Ph.D. from the Sorbonne in 1902. His dissertation is considered to be one of the best mathematics theses ever written. The first chapter develops his theory of measure; the second chapter develops his integral; the third chapter discusses length, area, and certain surfaces; the fourth chapter is on Plateau's problem about minimal surfaces. Many of the important properties of the Riemann integral were generalized by Lebesgue to his integral in the second chapter.

Probably the most notable exception to this is that what is now often referred to as the first version of fundamental theorem of calculus (that $\frac{d}{dx}\left(\int_a^x f(t)dt\right) = f(x)$ almost everywhere) does not appear. Lebesgue was aware that he wanted this statement and, in fact, was unable to prove it for his thesis. He was able to prove it later, and it appeared in print one year later [81]. Several other unresolved issues in his thesis were also resolved by Lebesgue himself during the two years following his thesis work.

At the end of the nineteenth century, only continuous functions could be dealt with in a satisfactory manner, and there was still much debate over what the definition of a function should be. By the end of the first decade of the twentieth century, the treatment of discontinuous functions was fully incorporated. This ten-year revolution, culminating in the modern theory of real functions, was led by Lebesgue. Lebesgue's ideas can be seen to be very strongly and most directly influenced by the works of René Baire, Emile Borel, and Camille Jordan.[1] Baire's work gave deep insights into the behavior of discontinuous functions, while the work of Borel and Jordan focused on measuring the size of sets. The ideas of Baire, Borel, and Jordan had, of course, interested others as well. In particular, the works of Giuseppe Vitali (1875–1932; Italy) and William Henry Young (1863–1942; England) should be noted in the context of the development of the measure and integral credited to Lebesgue (see [61]).

In 1904 Lebesgue published his book *Leçons sur l'intégration et la recherche des fonctions primitives*. This book reached a large number of readers, and it did not take long for Lebesgue's integral to become the integral of choice for most practitioners. It was taught to undergraduates as early as 1914, at the Rice Institute (now Rice University, in Texas). The Lebesgue integral has had remarkable success in applications, and its staying power is really because of these applications. Lebesgue himself applied his integral to problems having to do with trigonometric series, problems that had arisen in Fourier's work. As discussed in the opening paragraphs of Chapter 3, it is Riesz who deserves much credit for drawing attention to the importance of Lebesgue's ideas by showing their value for solving problems in the new field of functional analysis. Indeed, if it were not for Riesz's applications of Lebesgue's ideas, functional analysis would not have developed as it did and might look very different today. And as we have seen, the field of probability would not be the same without the notion of the Lebesgue integral.

By 1922, Lebesgue had published dozens of papers on set theory, integration, measure, trigonometric series, polynomial approximation, topology, and geometry. Over the next twenty years he continued to write, but the focus of his papers shifted toward the expository, often treating historical, philosophical, or pedagogical topics and reflecting his great interest and strong views on teaching.

Henri Lebesgue died on July 26, 1941, in Paris.

---

[1] Borel we have already encountered. Jordan was a French mathematician who lived from 1838 to 1922; the Jordan canonical form in matrix theory and the Jordan curve theorem in topology are two results named for him. Baire was also French, and lived from 1874 to 1932. One of Baire's results is the subject of Section 6.2.

3.3   Measurable and Lebesgue Integrable Functions on Euclidean Space

We will be using Euclidean space $\mathbb{R}^n$ as our measure space X, the Lebesgue measurable sets $\mathcal{M}$ as our σ-ring $\mathcal{R}$, and Lebesgue measure m as our measure μ. Everything that we say in this and in the next section for the triple $(\mathbb{R}^n, \mathcal{M}, m)$ can be said for the more general measure space $(X, \mathcal{R}, \mu)$. Note that the use of the phrase "measure space" introduced in the last sentence is different from the prior usage, when it was used to refer to X alone.

A function $f : \mathbb{R}^n \to \mathbb{R} \cup \{\pm\infty\}$ is called *measurable* if the set

$$\{x \mid f(x) > a\}$$

is measurable for each $a \in \mathbb{R}$.

Theorem 3.7.   *The following are equivalent statements:*

(a) $\{x \mid f(x) > a\}$ *is measurable for every* $a \in \mathbb{R}$.
(b) $\{x \mid f(x) \geq a\}$ *is measurable for every* $a \in \mathbb{R}$.
(c) $\{x \mid f(x) < a\}$ *is measurable for every* $a \in \mathbb{R}$.
(d) $\{x \mid f(x) \leq a\}$ *is measurable for every* $a \in \mathbb{R}$.

PROOF.   Theorem 3.6 shows that $\mathcal{M}$ is a σ-ring, and hence $A \in \mathcal{M}$ if and only if $A^c \in \mathcal{M}$. From this, (a) $\Leftrightarrow$ (d) and (b) $\Leftrightarrow$ (c) follow immediately.

That (a) implies (b) follows from Theorem 3.6 and the equalities

$$\{x \mid f(x) \geq a\} = \bigcap_{k=1}^{\infty} \left\{ x \mid f(x) > a - \frac{1}{k} \right\} = \left(\bigcup_{k=1}^{\infty} \left\{ x \mid f(x) \leq a - \frac{1}{k} \right\} \right)^c.$$

That (b) implies (a) follows from Theorem 3.6 and the equality

$$\{x \mid f(x) > a\} = \bigcup_{k=1}^{\infty} \left\{ x \mid f(x) \geq a + \frac{1}{k} \right\}. \qquad \square$$

Theorem 3.8.   *If f is measurable, then $|f|$ is measurable.*

PROOF.   Left as an Exercise 3.3.4. $\square$

Theorem 3.9.   *If f and g are measurable, then so are $f + g$, fg, f_+, and f_-. If $\{f_k\}_{k=1}^{\infty}$ is a sequence of measurable functions, then the four functions*

$$\left(\inf f_k \right)(x) = \inf\{ f_k(x) \mid 1 \leq k < \infty\},$$

$$\left(\sup f_k \right)(x) = \sup\{ f_k(x) \mid 1 \leq k < \infty\},$$

$$\left(\liminf f_k \right)(x) = \sup_{j \geq 1} \left(\inf_{k \geq j} f_k(x) \right),$$

$$\left(\limsup f_k \right)(x) = \inf_{j \geq 1} \left(\sup_{k \geq j} f_k(x) \right),$$

are each measurable.

PROOF. First, we prove that $f + g$ is measurable. Observe that

$$(f + g)(x) < a \Leftrightarrow f(x) < a - g(x)$$

and that this is true if and only if there exists a rational number r such that

$$f(x) < r < a - g(x).$$

Therefore,

$$\{x \,|(f + g)(x) < a\} = \bigcup_{r \in \mathbb{Q}} \left(\{x \,|f(x) < r\} \cap \{x \,|g(x) < a - r\} \right).$$

Since the right-hand side belongs to $\mathcal{M}$, so does the set on the left.

Next, we next prove that fg is measurable for the special case when $f = g$. We then use the first part of the theorem, the special case, and the so-called polarization identity

$$fg = \frac{1}{4}(f + g)^2 - \frac{1}{4}(f - g)^2,$$

to get the general case. The case $f = g$ is taken care of by noticing that

$$\{x \,|(ff)(x) > a\} = \{x \,|f(x) > \sqrt{a}\} \cup \{x \,|f(x) < -\sqrt{a}\}.$$

We next prove that $\sup f_k$ is measurable. For each $a \in \mathbb{R}$,

$$\{x \,|(\sup f_k)(x) > a\} = \bigcup_{k=1}^{\infty} \{x \,|f_k(x) > a\}.$$

Since each set in the union on the right side of the equation is in $\mathcal{M}$, so is the set on the left side.

The proof that $\inf f_k$ is measurable is similar to the argument for the supremum. Then $\limsup f_k$ and $\liminf f_k$ are measurable from these (applying the argument twice in succession). The facts that f_- and f_+ are measurable follow from the proof for $\sup f_k$, since $f_+ = \max\{f, 0\} = \sup\{f, 0\}$ and $f_- = \max\{-f, 0\} = \sup\{-f, 0\}$. $\square$

This last theorem can be interpreted as saying that the usual ways of combining functions preserve measurability. One way of combining functions is noticeably missing: composition. It is not the case that the composition of two measurable functions is again measurable. See, for example, [1] (page 57) or [70] (page 362) to see what can be said about the measurability of the composition of two functions.

A real-valued function with only a finite number of elements in its range is called a *simple function*. One type of simple function is the *characteristic function*, χ_E, of a set $E \subseteq \mathbb{R}^n$. This is defined by

$$\chi_E(x) = \begin{cases} 1 & \text{if } x \in E, \\ 0 & \text{if } x \notin E. \end{cases}$$

Every simple function can be written as a finite linear combination of characteristic functions. Specifically, if the range of the simple function s is $\{c_1, \ldots, c_N\}$, then

$$s(x) = \sum_{k=1}^{N} c_k \chi_{E_k}(x),$$

where $E_k = \{x : s(x) = c_k\}$. The function s is measurable if and only if each set E_k is in $\mathcal{M}$.

What might be more remarkable is that every function defined on $\mathbb{R}^n$ can be well approximated by simple functions. This is the thrust of the next theorem.

Theorem 3.10. *If f is a real-valued function defined on $\mathbb{R}^n$, then there exists a sequence $\{s_k\}_{k=1}^{\infty}$ of simple functions such that*

$$\lim_{k \to \infty} s_k(x) = f(x), \quad \text{for every } x \in \mathbb{R}^n.$$

Further, if f is measurable, then the s_k's may be chosen to be measurable simple functions. Finally, if $f \geq 0$, then $\{s_k\}_{k=1}^{\infty}$ may be chosen to satisfy $s_1 \leq s_2 \leq \cdots$.

PROOF.   We first consider the case that $f \geq 0$. In the general case, we use that $f = f_+ - f_-$ and apply the construction below to each of f_+ and f_-.

Fix $f \geq 0$ and a positive integer k; we start by defining the simple function s_k. Define

$$F_k = \{x \mid f(x) \geq k\},$$

and sets

$$E_j^k = \left\{x \mid \frac{j-1}{2^k} \leq f(x) < \frac{j}{2^k}\right\},$$

for each integer j, $1 \leq j \leq k2^k$. Then put

$$s_k(x) = k\chi_{F_k}(x) + \sum_{j=1}^{k2^k} \frac{j-1}{2^k} \chi_{E_j^k}(x).$$

It is left as an exercise to show that the sequence $\{s_k\}_{k=1}^{\infty}$ has all of the desired properties.   □

Let $E \in \mathcal{M}$. For a measurable simple function $s(x) = \sum_{k=1}^{N} c_k \chi_{E_k}(x)$, we define the *Lebesgue integral of s over E* by

$$\int_E s\, dm = \sum_{k=1}^{N} c_k m(E \cap E_k).$$

For a measurable, nonnegative function f we define the *Lebesgue integral of f over E* by

$$\int_E f\, dm = \sup\left\{\int_E s\, dm \mid 0 \leq s \leq f, \ s \text{ simple}\right\}.$$

Note that $\int_E f\,dm$ may be infinite. Now let f be an arbitrary (not necessarily nonnegative) measurable function. We say that f is *integrable* if both

$$\int_E f_+dm \quad \text{and} \quad \int_E f_-dm$$

are finite and define

$$\int_E f\,dm = \int_E f_+dm - \int_E f_-dm.$$

Integrability is really a statement about *absolute* integrability, as will be seen in the exercises (Exercise 3.3.7). This sometimes causes confusion. The integral

$$\int_E f\,dm$$

is referred to as the *integral of f, with respect to the measure m, over E*. This terminology opens the door for integrating with respect to other measures. We will discuss other measures in the last section of this chapter. We let

$$\mathcal{L}(\mathbb{R}^n, m) \quad \text{or} \quad \mathcal{L}(\mathbb{R}^n)$$

denote the collection of all functions that are integrable with respect to Lebesgue measure m over $\mathbb{R}^n$. This collection forms a real linear space. This fact and other useful properties of the integral are listed in the following theorem.

Theorem 3.11. *The Lebesgue integral enjoys several properties.*

(a) *The integral is linear. That is,*

$$\int_E cf\,dm = c\int_E f\,dm \quad \text{and} \quad \int_E (f+g)dm = \int_E f\,dm + \int_E g\,dm$$

whenever $f, g \in \mathcal{L}(\mathbb{R}^n)$, $c \in \mathbb{R}$, and $E \in \mathcal{M}$.

(b) *The integral is monotone. That is,*

$$\int_E f\,dm \le \int_E g\,dm$$

whenever $f, g \in \mathcal{L}(\mathbb{R}^n)$, $f(x) \le g(x)$ for all $x \in E$.

(c) *For every $f \in \mathcal{L}(\mathbb{R}^n)$, we have $|f| \in \mathcal{L}(\mathbb{R}^n)$ and*

$$\left| \int_E f\,dm \right| \le \int_E |f|\,dm.$$

(d) *For every $f \in \mathcal{L}(\mathbb{R}^n)$, we have $\int_E f\,dm = 0$ for every measurable set E of measure zero. From this it follows that*

$$\int_A f\,dm = \int_B f\,dm$$

whenever A and B are measurable sets, $B \subseteq A$, and $m(A \setminus B) = 0$.

PROOF.   A proof of the first part of (a) is straightforward, as is a proof of (b); (c) follows from (b); (d) is straightforward as well. The second part of (a), which certainly should hold if there is any justice in the world, is more subtle than it appears; we will prove it using Lebesgue's monotone convergence theorem (Theorem 3.13). Proofs of the first part of (a), the second part of (a) for simple functions, (b), (c), and (d) are asked for in Exercise 3.3.8.   □

In general, if a property P holds on a set A except possibly at each point of some subset of A that can be contained in a measurable set of measure zero, then we say that the property P holds on A *almost everywhere*, or *for almost all $x \in A$*. In light of (d), the phrase "$f(x) \leq g(x)$ for all $x \in E$" in (b) can be replaced by the phrase "$f(x) \leq g(x)$ for almost all $x \in E$." From now on in this chapter expressions such as $f \leq g$, $f = g$, etc., should be interpreted as $f(x) \leq g(x)$ for almost all x, $f(x) = g(x)$ for almost all x, etc.

We end this section with a further property of the integral. We will use this result to prove Lebesgue's monotone convergence theorem.

Theorem 3.12. *Assume that $f \geq 0$ is measurable, and that $A_1, A_2, \ldots \in \mathcal{M}$ are pairwise disjoint. Then,*

$$\int_{\bigcup_{k=1}^{\infty} A_k} f \, dm = \sum_{k=1}^{\infty} \left(\int_{A_k} f \, dm \right).$$

PROOF.   We first consider the case that $f = \chi_E$ for some $E \in \mathcal{M}$. By the countable additivity of m, we have

$$\int_A f \, dm = m(A \cap E) = m \left(\bigcup_{k=1}^{\infty} (A_k \cap E) \right)$$

$$= \sum_{k=1}^{\infty} m(A_k \cap E) = \sum_{k=1}^{\infty} \left(\int_{A_k} f \, dm \right).$$

The next case, that f is simple, follows from this first case by the way that we define the integral for simple functions.

Finally, we consider an arbitrary measurable $f \geq 0$. Let $\epsilon > 0$ and choose a simple function s such that $s \leq f$ and

$$\int_A f \leq \left(\int_A s \right) + \epsilon.$$

The right-hand side of this inequality is equal to

$$\sum_{k=1}^{\infty} \left(\int_{A_k} s \, dm \right) + \epsilon \leq \sum_{k=1}^{\infty} \left(\int_{A_k} f \, dm \right) + \epsilon.$$

Thus,

$$\int_A f \leq \sum_{k=1}^{\infty} \left(\int_{A_k} f \, dm \right).$$

The proof will be complete when we show that also

$$\int_A f \geq \sum_{k=1}^{\infty} \left(\int_{A_k} f\,dm \right).$$

We first consider two disjoint sets $A_1, A_2 \in \mathcal{M}$ and choose two simple functions s_1, s_2 such that $0 \leq s_k \leq f$ and

$$\int_{A_k} s_n dm \geq \left(\int_{A_k} f\,dm \right) - \frac{\epsilon}{2}, \quad k = 1, 2.$$

Set $s = \max\{s_1, s_2\}$. Then s is simple and $0 \leq s \leq f$. Also,

$$\int_{A_k} s\,dm \geq \left(\int_{A_k} f\,dm \right) - \frac{\epsilon}{2}, \quad k = 1, 2.$$

Therefore,

$$\int_{A_1} s\,dm + \int_{A_2} s\,dm \geq \int_{A_1} f\,dm + \int_{A_2} f\,dm - \epsilon.$$

Put $A = A_1 \cup A_2$. By the first part of the theorem,

$$\int_A s\,dm \geq \int_{A_1} f\,dm + \int_{A_2} f\,dm - \epsilon.$$

By monotonicity,

$$\int_A f\,dm \geq \int_A s\,dm,$$

and so

$$\int_A f\,dm \geq \int_{A_1} f\,dm + \int_{A_2} f\,dm - \epsilon.$$

Since ϵ was arbitrary, we have shown that

$$\int_A f\,dm \geq \int_{A_1} f\,dm + \int_{A_2} f\,dm.$$

We now use induction to show that

$$\int_A f\,dm \geq \sum_{k=1}^{N} \left(\int_{A_k} f\,dm \right), \quad N = 1, 2, \ldots.$$

Finally, we return to the general case $A = \bigcup_{k=1}^{\infty} A_k$. For any positive integer N, the preceding inductive argument shows that

$$\int_A f\,dm \geq \int_{A_1 \cup A_2 \cup \cdots \cup A_N} f\,dm \geq \sum_{k=1}^{N} \left(\int_{A_k} f\,dm \right).$$

Since this holds for each N, we are done.   □

As you are asked to prove in Exercise 3.3.1, continuous functions are always measurable. Also, many continuous functions are integrable; for example, all continuous, bounded functions that vanish outside of some finite interval are integrable. How *discontinuous* can an element of $\mathcal{L}(\mathbb{R}^n)$ be? We know that we can take any continuous integrable function, alter its value on a set M of measure zero, and still have an integrable function. For example, the set M can be taken to be a countable dense subset of $\mathbb{R}^n$. Nonetheless, the continuous functions with "compact support" are dense in $\mathcal{L}(\mathbb{R}^n)$ (see Exercise 3.6.8 to see precisely what is meant by compact support).

3.4   The Convergence Theorems

In this section we shall see three theorems about how the Lebesgue integral behaves with respect to limit operations. The properties revealed in these theorems are what distinguish the Lebesgue integral from competitor integrals.

Theorem 3.13 (Lebesgue's Monotone Convergence Theorem). *Suppose that $A \in \mathcal{M}$ and that $\{f_k\}_{k=1}^{\infty}$ is a sequence of measurable functions such that*

$$0 \le f_1(x) \le f_2(x) \le \cdots \quad \text{for almost all } x \in A.$$

Let f be defined to be the pointwise limit, $f(x) = \lim_{k\to\infty} f_k(x)$, of this sequence. Then f is integrable and

$$\lim_{k\to\infty} \left(\int_A f_k dm \right) = \int_A f\, dm.$$

PROOF.   We have

$$0 \le f_1(x) \le f_2(x) \le \cdots \le f(x) = \lim_{k\to\infty} f_k(x) \quad \text{for almost all } x \in A.$$

By monotonicity, we get

$$\int_A f_1 dm \le \int_A f_2 dm \le \cdots \le \int_A f\, dm.$$

Thus $\{\int_A f_k dm\}_{k=1}^{\infty}$ is a bounded and nondecreasing sequence of real numbers, and hence must converge to some real number L. Note that $L \le \int_A f\, dm$; we aim to show that $L \ge \int_A f\, dm$ also. To do this we choose a number $\delta \in (0, 1)$ and a simple function s satisfying $0 \le s(x) \le f(x)$ for almost all $x \in A$. Define

$$A_k = \{x \in A \mid f_k(x) \ge \delta s(x)\}.$$

Then

$$A_1 \subseteq A_2 \subseteq A_3 \subseteq \cdots$$

and

$$A = \bigcup_{k=1}^{\infty} A_k.$$

For each positive integer k, we have

$$L = \lim_{k \to \infty} \left(\int_A f_k dm \right) \geq \int_A f_k dm \geq \int_{A_k} f_k dm \geq \delta \int_{A_k} s dm.$$

Therefore,

$$L \geq \delta \cdot \lim_{k \to \infty} \left(\int_{A_k} s dm \right).$$

We claim that

$$\lim_{k \to \infty} \left(\int_{A_k} s dm \right) = \int_A s dm. \tag{†}$$

Given that this equality holds, we obtain

$$L \geq \delta \cdot \int_A s dm.$$

Since δ was arbitrary,

$$L \geq \int_A s dm.$$

Taking the supremum over all such simple functions now yields

$$L \geq \int_A f dm.$$

This is what we wanted to prove.

To see (†), put $E_1 = A_1$, and $E_k = A_k \setminus A_{k-1}$. Then

$$A = \bigcup_{k=1}^{\infty} E_k, \qquad A_k = \bigcup_{j=1}^{k} E_j,$$

and the E_k's are pairwise disjoint. Theorem 3.12 then implies that

$$\int_A f dm = \sum_{k=1}^{\infty} \left(\int_{E_k} f dm \right),$$

which, by definition, is equal to

$$\lim_{k \to \infty} \left(\sum_{j=1}^{k} \left(\int_{E_j} f dm \right) \right) = \lim_{k \to \infty} \left(\int_{A_k} f dm \right). \qquad \square$$

Before moving on to the next "convergence theorem" we fulfill our promise made in the previous section and use Lebesgue's monotone convergence theorem to prove the second part of Theorem 3.11(a). Specifically,

$$\int_E (f + g) dm = \int_E f dm + \int_E g dm$$

whenever $f, g \in \mathcal{L}(\mathbb{R}^n)$ and $E \in \mathcal{M}$. In Exercise 3.3.8(b) you are asked to prove the result in the case that f and g are simple functions. Since $\int_E f dm$ is defined by the difference $\int_E f_+ dm - \int_E f_- dm$, we may assume that f is nonnegative (almost everywhere). Likewise for g. We first appeal to Theorem 3.10 to get sequences of nonnegative, measurable, simple functions $\{s_k\}_{k=1}^\infty$ and $\{t_k\}_{k=1}^\infty$ satisfying

$$\lim_{k\to\infty} s_k(x) = f(x), \qquad \lim_{k\to\infty} t_k(x) = g(x)$$

almost everywhere. Combining the monotone convergence theorem and the result for simple functions from the exercises, we see that

$$\begin{aligned}
\int_E (f + g)dm &= \lim_{k\to\infty} \int_E (s_k + t_k)dm \\
&= \lim_{k\to\infty} \int_E s_k dm + \lim_{k\to\infty} \int_E t_k dm \\
&= \int_E f dm + \int_E g dm.
\end{aligned}$$

The next result was proved by Pierre Fatou (1878–1929; France) in his 1906 doctoral dissertation. Fatou was also an astronomer. He studied twin stars and proved a conjecture of Gauss's on planetary orbits.

Theorem 3.14 (Fatou's Lemma). *Assume that $A \in \mathcal{M}$. Let $\{f_k\}_{k=1}^\infty$ be a sequence of nonnegative measurable functions and let $f = \liminf_{k\to\infty} f_k$ on A. Then*

$$\int_A f dm \le \liminf_{k\to\infty} \left(\int_A f_k dm \right).$$

PROOF.   For each positive integer j, define a function g_j by

$$g_j(x) = \inf_{k\ge j} f_k(x),$$

and a number a_j by

$$a_j = \inf_{k\ge j} \left(\int_A f_k dm \right).$$

Theorem 3.9 shows that each g_j is measurable, and clearly $\sup_{j\ge 1} g_j = f$. Since

$$0 \le g_1(x) \le g_2(x) \le \cdots,$$

we have that

$$\lim_{j\to\infty} g_j = \sup_{j\ge 1} g_j = f.$$

Since

$$0 \le a_1(x) \le a_2(x) \le \cdots,$$

we have that

$$\lim_{j\to\infty} a_j = \sup_{j\ge 1} a_j = \liminf_{k\to\infty} \left(\int_A f_k dm \right).$$

Observe that $g_j(x) \leq f_k(x)$ for each pair of positive integers j, k with $k \geq j$. Thus,

$$\int_A g_j dm \leq a_j,$$

for each positive integer j. The monotone convergence theorem now implies that

$$\int_A f dm \leq \lim_{j \to \infty} \left(\int_A g_j dm \right) \leq \lim_{j \to \infty} a_j = \liminf_{k \to \infty} \left(\int_A f_k dm \right). \qquad \square$$

The following is one of the best results for telling us when we may conclude that

$$\lim_{k \to \infty} \left(\int_A f_k dm \right) = \int_A \left(\lim_{k \to \infty} f_k \right) dm.$$

Theorem 3.15 (Lebesgue's Dominated Convergence Theorem). *Assume that $A \in \mathcal{M}$. Let $\{f_k\}_{k=1}^{\infty}$ be a sequence of measurable functions, and put $f(x) = \lim_{k \to \infty} f_k(x)$. Further, assume that there exists a function $g \in \mathcal{L}(\mathbb{R}^n)$ such that $|f_k(x)| \leq g(x)$ for almost all $x \in A$ and each positive integer k. Then we may conclude that*

$$\lim_{k \to \infty} \left(\int_A f_k dm \right) = \int_A f dm.$$

PROOF.   Begin by noticing that for each k, $(f_k)_+ \leq g$, and $(f_k)_- \leq g$ and thus each f_k is in $\mathcal{L}(\mathbb{R}^n)$.

We first want to see that $|f|$ is in $\mathcal{L}(\mathbb{R}^n)$. This follows from Fatou's lemma:

$$\int_A |f| dm = \int_A \left(\lim_{k \to \infty} |f_k| \right) dm = \int_A \left(\liminf_{k \to \infty} |f_k| \right) dm$$

$$\leq \liminf_{k \to \infty} \left(\int_A |f_k| dm \right) \leq \int_A g dm.$$

Since each $f_k + g$ is a nonnegative function, Fatou's lemma shows that

$$\int_A f dm + \int_A g dm = \int_A (f + g) dm$$

$$= \int_A \left(\liminf_{k \to \infty} (f_k + g) \right) dm \leq \liminf_{k \to \infty} \left(\int_A (f_k + g) dm \right).$$

Because the integral and the processes of taking infima and suprema are additive, the expression on the right becomes

$$\liminf_{k \to \infty} \left(\int_A f_k dm \right) + \int_A g dm.$$

Combining these yields

$$\int_A f dm \leq \liminf_{k \to \infty} \left(\int_A f_k dm \right).$$

Since each $g - f_k$ is a nonnegative function, we can repeat this argument and get

$$\int_A f\,dm \geq \limsup_{k\to\infty}\left(\int_A f_k\,dm\right).$$

Combining these last two inequalities yields the desired result.   □

3.5 Comparison of the Lebesgue Integral with the Riemann Integral

Lebesgue developed his integral in an effort to perfect the integral of Riemann. The main goal of this section is to show that the Riemann integrable functions form a proper subcollection of the Lebesgue integrable functions. In this section we will give several results without including proofs of them. Proofs can be found in any of the books on integration mentioned in the bibliography. *The discussion in this section is limited to integration on $\mathbb{R}$.*

We begin with a brief review of the definition of the Riemann integral. We assume that the reader is familiar with the Riemann integral and its properties and include this material as a reminder and also to establish notation.

We consider a bounded, real-valued function f defined on the closed and bounded interval $[a, b]$. A collection of points $P = \{x_0, x_1, \ldots, x_n\}$ is called a *partition* of $[a, b]$ if

$$a = x_0 < x_1 < \cdots < x_n = b.$$

The length of the longest subinterval $[x_{k-1}, x_k]$ is called the *mesh* of the partition P. Set, for $k = 1, 2, \ldots, n$,

$$m_k = \inf\{f(x)\,|\,x \in [x_{k-1}, x_k]\} \qquad \text{and} \qquad M_k = \sup\{f(x)\,|\,x \in [x_{k-1}, x_k]\}.$$

The *lower Riemann sum* $L(f, P)$ of f corresponding to the partition P is given by

$$L(f, P) = \sum_{k=1}^{n} m_k(x_k - x_{k-1}).$$

and the *upper Riemann sum* $U(f, P)$ of f corresponding to the partition P is given by

$$U(f, P) = \sum_{k=1}^{n} M_k(x_k - x_{k-1}).$$

The *lower Riemann integral* of f is defined by

$$L(f) = \sup\{L(f, P)\,|\,P \text{ is a partition of } [a, b]\},$$

and the *upper Riemann integral* of f is defined by

$$U(f) = \inf\{U(f, P)\,|\,P \text{ is a partition of } [a, b]\}.$$

Finally, a bounded, real-valued function f defined on $[a, b]$ is called *Riemann integrable* if $L(f) = U(f)$. In this case, their common value is denoted by

$$\int_a^b f(x)dx.$$

In the preceding paragraph, that the sets used above to define the lower and upper integrals do indeed have upper and lower bounds, respectively, is something one must prove. There is one result about Riemann integrals that we will use in this section: If $\{P_n\}_{n=1}^{\infty}$ is any sequence of partitions of $[a, b]$ such that the meshes of the P_n's converge to zero as $n \to \infty$, then

$$\lim_{n \to \infty} L(f, P_n) = \lim_{n \to \infty} U(f, P_n) = \int_a^b f(x)dx.$$

The Riemann and Lebesgue integrals of a nonnegative real-valued function can be interpreted in terms of area, as you should recall. In a naive way, the difference between the two integrals (on $\mathbb{R}$) can be visualized by noting that the Riemann approach subdivides the domain of the integrand, while the Lebesgue approach subdivides the range.

Recall that Lebesgue was trying, among other things, to increase the number of integrable functions. Our next theorem shows that each Riemann integrable function is Lebesgue integrable (showing that Lebesgue's collection contains Riemann's collection). If we consider the characteristic function of the rational numbers in the interval $[0, 1]$, we get a function with $L(f, P) = 0$ and $U(f, P) = 1$ for each partition of the unit interval. Thus, we have a function that is not Riemann integrable. However, this function is Lebesgue integrable, as you are asked to prove in Exercise 3.3.3. This together with Lemma 3.16 shows that Lebesgue's collection is, in fact, larger than Riemann's. Lebesgue was successful in enlarging the class of integrable functions.

Theorem 3.16. *If f is Riemann integrable on $[a, b]$, then f is Lebesgue integrable on $[a, b]$ and*

$$\int_{[a,b]} f\,dm = \int_a^b f(x)dx.$$

(Recall that the integral on the left is the Lebesgue integral, and the integral on the right is the Riemann integral.)

PROOF. For each positive integer n, partition $[a, b]$ into 2^n subintervals each of length $\frac{b-a}{2^n}$. Let P_n denote this partition and $a = x_0 < x_1 < \cdots < x_n = b$ denote the points of P_n. Define

$$g_n(x) = \sum_{k=1}^{2^n} m_k \chi_{[x_{k-1}, x_k)}(x) \qquad \text{and} \qquad h_n(x) = \sum_{k=1}^{2^n} M_k \chi_{[x_{k-1}, x_k)}(x).$$

Then $\{g_n\}_{n=1}^{\infty}$ is an increasing sequence, and $\{h_n\}_{n=1}^{\infty}$ is a decreasing sequence. Put

$$g(x) = \lim_{n \to \infty} g_n(x) \qquad \text{and} \qquad h(x) = \lim_{n \to \infty} h_n(x).$$

Then g and h are Lebesgue integrable functions that satisfy

$$g(x) \leq f(x) \leq h(x)$$

for almost all x in $[a, b]$. Also,

$$\int_{[a,b]} g\,dm = L(f, P_n) \qquad \text{and} \qquad \int_{[a,b]} h\,dm = U(f, P_n).$$

Now, $h_n(x) - g_n(x) \geq 0$ for almost all x and

$$\lim_{n \to \infty} \left(h_n(x) - g_n(x) \right) = h(x) - g(x).$$

Therefore,

$$0 \leq \int_{[a,b]} (h - g)\,dm = \lim_{n \to \infty} \left(\int_{[a,b]} (h_n - g_n)\,dm \right)$$

$$= \lim_{n \to \infty} \int_{[a,b]} h_n\,dm - \lim_{n \to \infty} \int_{[a,b]} g_n\,dm$$

$$= \lim_{n \to \infty} U(f, P_n) - \lim_{n \to \infty} L(f, P_n) = 0.$$

The first equality follows from Lebesgue's monotone convergence theorem, and the last from the result about Riemann integrals referred to in the paragraph preceding this theorem. It now follows that $g = h$ almost everywhere. Thus f is Lebesgue integrable, and

$$\int_{[a,b]} f\,dm = \lim_{n \to \infty} \int_{[a,b]} g_n\,dm = \lim_{n \to \infty} L(f, P_n) = \int_a^b f(x)\,dx. \qquad \square$$

A Riemann integrable function must, by definition, be bounded. Which bounded functions are Riemann integrable? Different characterizations exist, and perhaps most notable is Riemann's own characterization: A bounded function $f : [a, b] \to \mathbb{R}$ is Riemann integrable if and only if for every $\epsilon > 0$ there exists a partition P of $[a, b]$ such that $U(f, P) - L(f, P) < \epsilon$. Lebesgue gave a characterization in terms of measure. Specifically, a bounded function $f : [a, b] \to \mathbb{R}$ is Riemann integrable if and only if it is continuous almost everywhere.

It would be remiss not to mention that a version of the fundamental theorem of calculus can be given for the Lebesgue integral. In it, modifications are made to allow for the possibility of bad behavior on a set of measure zero. If f is Lebesgue integrable on $[a, b]$ and we define F by

$$F(x) = \int_{[a,x]} f\,dm,$$

then we cannot conclude that $F'(x) = f(x)$ for every value of x in $[a, b]$ (nor even that F is differentiable everywhere), but we can conclude that F is differentiable almost everywhere and that $F'(x) = f(x)$ for almost every value of x in $[a, b]$.

Finally, let us consider once again the characteristic function of the rationals χ_Q. This function is not Riemann integrable on the interval $[0, 1]$ but is the pointwise limit of the sequence of Riemann integrable functions

$$f_n(x) = \begin{cases} 1 & \text{if } x \in \{r_1, \dots, r_n\}, \\ 0 & \text{otherwise,} \end{cases}$$

where $r_1, r_2, \dots$ is an enumeration of the rational numbers in the unit interval. In general, we can conclude that the limit of a sequence of Riemann integrable functions is again Riemann integrable if the convergence of the sequence is uniform (though uniform convergence is not necessary). The convergence theorems of Lebesgue (given in the previous section) show that the requirement of uniform convergence may be greatly relaxed to pointwise convergence if other, less restrictive, requirements are imposed on the sequence of functions when the Lebesgue integral is used in place of the Riemann integral. The fact that pointwise limit and the Lebesgue integral may be interchanged is a key property that makes the Lebesgue integral more useful than Riemann's integral.

3.6   General Measures and the Lebesgue L^p-spaces: The Importance of Lebesgue's Ideas in Functional Analysis

At the beginning of the second section of this chapter, we alluded to arbitrary measure spaces $(X, \mathcal{R}, \mu)$. We now give a discussion of these. Recall that a measure space consists of three things:

(i) a nonempty set X;

(ii) a σ-ring $\mathcal{R}$ of subsets of X;

(iii) a function μ defined on $\mathcal{R}$ satisfying

(a) $0 \le \mu(A) \le \infty$ for all $A \in \mathcal{R}$,

(b) $\mu\left(\bigcup_{n=1}^{\infty} A_n\right) = \sum_{n=1}^{\infty} \mu(A_n)$ whenever $A_1, A_2, \dots \in \mathcal{R}$ satisfy $A_n \cap A_m = \emptyset$, $n \ne m$.

In the preceding sections we have constructed and studied Lebesgue measure on Euclidean space. This is certainly the most important example for our purposes. In this section we meet a few other examples of measure spaces and then introduce, for each measure space $(X, \mathcal{R}, \mu)$, the linear space $L^p(X, \mu)$. As stated at the beginning of Section 3, all the results of that section and of Section 4 that are proved for $(\mathbb{R}^n, \mathcal{M}, m)$ hold for any general measure space $(X, \mathcal{R}, \mu)$. We will use these generalizations freely in this section.

We first give a list of some examples of triples $(X, \mathcal{R}, \mu)$.

EXAMPLE 1.   Let $X = \mathbb{R}^n$, $\mathcal{R} = \mathcal{M}$, $\mu = m$; this is the example we have been considering.

EXAMPLE 2.   Let $X = \mathbb{R}^n$, $\mathcal{R} = \mathcal{B}$, $\mu = m$. Here, $\mathcal{B}$ is defined to be the smallest σ-ring containing all open subsets of $\mathbb{R}^n$. The elements of $\mathcal{B}$ are called the *Borel sets* of $\mathbb{R}^n$. Since all open sets are measurable, the Borel sets form a subring of the measurable sets. However, not all measurable sets are Borel sets, and so the two collections are different. To construct a measurable set that is not a Borel set, one can make a "Cantor-like" construction (see, for example, [71] page 110). The following, however, is true: If A is measurable, then A can be written as $(A \setminus B) \cup B$ for some Borel set $B \subseteq A$ satisfying $m(A \setminus B) = 0$. The Borel sets are often favored as the underlying ring because although the ring contains fewer sets, the elements of it can be described more readily than the Lebesgue measurable sets can be.

EXAMPLE 3.   Let $X = \mathbb{R}^n$, $\mathcal{R} = \mathcal{M}$. To define the measure, we consider any nondecreasing, continuous function $f : \mathbb{R} \to \mathbb{R}$ and put

$$\mu([a, b]) = f(b) - f(a).$$

Then μ can be extended to all of $\mathcal{M}$ in the same way that Lebesgue measure was. Indeed, this measure μ reduces to Lebesgue measure in the case $f(x) = x$. (The scope of this example can be increased greatly.)

EXAMPLE 4.   Let X be any set, $\mathcal{R} = 2^X$, and let μ be *counting measure*:

$$\mu(A) = \begin{cases} |A| & \text{if } A \text{ is finite,} \\ \infty & \text{if } A \text{ is infinite,} \end{cases}$$

where $|A|$ denotes the number of elements in A. This measure might seem a bit simplistic. It is, but it plays an important role in the L^p-theory.

We generate further examples by restricting the space X:

EXAMPLE 5. $X = [0, 1]$ (the unit interval), $\mathcal{R} = \{S \subseteq [0, 1] \,\big|\, S \in \mathcal{M}\}$, $\mu = m$.

EXAMPLE 6.   Let X be any uncountable set,

$$\mathcal{R} = \{A \in 2^X \,\big|\, A \text{ is countable or } X \setminus A \text{ is countable}\},$$

and

$$\mu(A) = \begin{cases} 0 & \text{if } A \text{ is countable,} \\ 1 & \text{if } X \setminus A \text{ is countable.} \end{cases}$$

EXAMPLE 7.   Let X be any finite or countable set and $\mathcal{R} = 2^X$. Write $X = \{x_1, x_2, \ldots\}$. Let p_i be a positive number corresponding to each x_i, and assume that $\sum p_i = 1$. Define μ by

$$\mu(A) = \sum_{x_i \in A} p_i.$$

Examples 5, 6, and 7 share the property that the measure of the entire space is 1. Any measure with this property is called a *probability measure*. Our first example (Example 5) of such a measure plays a critical role in abstract probability, as indicated in the first section of this chapter. Our second example (Example 6) is rather silly but, nonetheless, provides another example. Example 7 provides a model for discrete probability theory.

We begin by fixing a measure space $(X, \mathcal{R}, \mu)$ and a real number $1 \leq p < \infty$. We will consider $0 < p < 1$ in Exercise 3.6.2 and the important case $p = \infty$ later in this section. Notice that if $f : X \rightarrow \mathbb{R} \cup \{\pm\infty\}$ is a measurable function, then $|f|^p$ is also measurable.

Define, for a measure space $(X, \mathcal{R}, \mu)$ and a real number $1 \leq p < \infty$, the *Lebesgue space* $L^p(X, \mu)$ (or just $L^p(\mu)$, or even just L^p if the measure is clear from context) to be the collection of all μ-measurable functions such that

$$\int_X |f|^p d\mu < \infty.$$

We define the *p-norm* of an element $f \in L^p(\mu)$ to be the number

$$\|f\|_p = \left(\int_X |f|^p d\mu \right)^{\frac{1}{p}}.$$

The theory of L^p-spaces was developed by F. Riesz in 1910 [105]. In that article he introduced these spaces for Lebesgue measure on measurable subsets of $\mathbb{R}^n$, proved the Hölder and Minkowski inequalities (see our Theorems 3.18 and 3.19) in this setting, and showed that the step functions are dense in these spaces (our Theorem 3.22). He also showed that these spaces are norm complete (our Theorem 3.21). For $p = 2$ this had already been shown by E. Fischer [42]. Riesz's 1910 paper was a remarkable achievement, and remains one of the most important papers ever published in the field.

We now make two critical remarks regarding these definitions. First, we will want to prove that $\|f\|_p$ defines a norm, and in particular that $\|f\|_p = 0$ if and only if "$f = 0$." The equality in quotation marks where we must be careful. We know that the integral of a function will be zero as long as the function is equal, almost everywhere, to zero. In fact, $L^p(\mu)$ really consists of equivalence classes of functions rather than of functions, where

$$f \sim g \text{ if and only if } f(x) = g(x) \text{ almost everywhere.}$$

We will rarely mention this distinction, but it is important, and you should do your best to understand this point. As our second remark about this definition we point out that we are interested in being able to consider *complex*valued functions. Until now however, integration has been discussed only for real-valued functions. A function $f : X \rightarrow \mathbb{C}$ is called measurable if both re(f) and im(f) are measurable real-valued functions. In this case, we define the integral of f by

$$\int_X f d\mu = \int_X \text{re}(f) d\mu + i \int_X \text{im}(f) d\mu.$$

We will also use $L^p(\mu)$ to include complex-valued functions.

In summary, $L^p(\mu)$ denotes the set of all equivalence classes of complex-valued functions f defined on X satisfying

$$\int_X |f|^p d\mu < \infty.$$

We write f to stand for the equivalence class of all functions that are equal to f almost everywhere.

Theorem 3.17. *For $1 \leq p < \infty$, $L^p(\mu)$ is a linear space.*

PROOF. This is easy to see. It is trivial to show that $\lambda f \in L^p(\mu)$ whenever $f \in L^p(\mu)$ and $\alpha \in \mathbb{C}$. To see that $f + g \in L^p(\mu)$ whenever both f and g are in $L^p(\mu)$, we use the inequality

$$|f + g|^p \leq 2^p(|f|^p + |g|^p). \qquad \square$$

Notice that for any $1 < p < \infty$ there is a unique number q such that

$$\frac{1}{p} + \frac{1}{q} = 1.$$

If $p = 1$, we define $q = \infty$; if $p = \infty$, we define $q = 1$. This number q is sometimes called the *Hölder conjugate* of p. Note that the Hölder conjugate of 2 is itself, and that this is the only number that is its own Hölder conjugate.

Let y be a fixed nonnegative number, and $1 < p < \infty$. The maximum of the function $f(x) = xy - \frac{x^p}{p}$ occurs at $x = y^{\frac{1}{p-1}}$ and thus $f(x) \leq f(y^{\frac{1}{p-1}})$ for all nonnegative numbers x. This inequality can be rearranged to yield

$$xy \leq \frac{x^p}{p} + \frac{y^q}{q}$$

for all nonnegative numbers x and y (Exercise 3.6.3). We use this to prove our next result.

Theorem 3.18 (Hölder's Inequality). *Assume that $1 < p < \infty$ and $1 < q < \infty$ are Hölder conjugates, and that $f \in L^p$ and $g \in L^q$. Then $fg \in L^1$ and*

$$\|fg\|_1 \leq \|f\|_p \|g\|_q.$$

PROOF. If $f = 0$ or $g = 0$ (recall that we mean here that $f = 0$ almost everywhere or $g = 0$ almost everywhere), then the result is trivial. So, we assume that $\|f\|_p > 0$ and $\|g\|_q > 0$. The discussion preceding this theorem shows that

$$\frac{|f(x)g(x)|}{\|f\|_p \|g\|_q} \leq \frac{|f(x)|^p}{p(\|f\|_p)^p} + \frac{|g(x)|^q}{q(\|g\|_q)^q}.$$

Integrating both sides of this yields the desired result. $\qquad \square$

The German mathematician Otto Ludwig Hölder (1859–1937) worked mostly in group theory. However, he did work in analysis on the convergence of Fourier series (see Chapter 4). He proved his inequality in 1884.

Theorem 3.19. *For* $1 \le p < \infty$, $L^p(\mu)$ *is a normed linear space, with norm given by*

$$\|f\|_p = \left(\int_X |f|^p d\mu \right)^{\frac{1}{p}}.$$

PROOF.   It is straightforward to see that $\|f\|_p \ge 0$ for all $f \in L^p(\mu)$, that equality holds if and only if $f = 0$ almost everywhere, and that $\|\lambda f\|_p = |\lambda| \|f\|_p$ for $\lambda \in \mathbb{C}$. We concentrate our efforts on verifying the triangle inequality:

$$\|f + g\|_p \le \|f\|_p + \|g\|_p, \quad \text{for } f, g \in L^p(\mu).$$

If $p = 1$, this follows from Theorem 3.11. In the case $1 < p < \infty$, the result is nontrivial and is called *Minkowski's inequality*. If $\|f + g\|_p = 0$, there is nothing to prove, so we assume that this is not the case. We first note that

$$\| |f + g|^{p-1} \|_q = \left(\|f + g\|_{(p-1)q} \right)^{p-1} = \left(\|f + g\|_p \right)^{p-1}.$$

(This is left as Exercise 3.6.4.) Hölder's inequality then implies

$$
\begin{aligned}
1 &= \frac{1}{(\|f + g\|_p)^p} \left(\int_X |f + g|^p d\mu \right) \\
&= \frac{1}{(\|f + g\|_p)^p} \left(\int_X |f + g| \cdot |f + g|^{p-1} d\mu \right) \\
&\le \frac{1}{(\|f + g\|_p)^p} \left(\int_X |f| \cdot |f + g|^{p-1} d\mu + \int_X |g| \cdot |f + g|^{p-1} d\mu \right) \\
&\le \frac{1}{(\|f + g\|_p)^p} \left(\|f\|_p \cdot \| |f + g|^{p-1} \|_q + \|g\|_p \cdot \| |f + g|^{p-1} \|_q \right) \\
&= \frac{1}{\|f + g\|_p} \left(\|f\|_p + \|g\|_p \right).
\end{aligned}
$$

□

Let $(X, \mathcal{R}, \mu)$ be a measure space. A measurable function f is said to be *essentially bounded* if there exists a nonnegative real number M and a measurable set A of measure zero such that

$$|f(x)| \le M, \quad \text{for all } x \in X \setminus A.$$

Then $L^\infty(X, \mu)$ (or $L^\infty(X)$ or even just L^∞) is defined to be the set of all essentially bounded measurable functions. We define $\|f\|_\infty$ for these functions by

$$\|f\|_\infty = \inf\{M\},$$

where the infimum is taken over all M that provide a bound in the definition of f being essentially bounded.

It is straightforward to verify that L^∞ is a linear space, and that $\| \cdot \|_\infty$ satisfies the properties to make L^∞ into a normed linear space. The space L^∞ (for Lebesgue measure on an interval of $\mathbb{R}$) was introduced by Hugo Steinhaus (1887–1972; Poland) in [116]. We have read a bit about him in Banach's biography. Steinhaus made many contributions to probability, functional analysis, and game theory. In 1923, Steinhaus published the first truly mathematical treatment of coin tossing based on measure theory. He is also known for his very popular books *Mathematical Snapshots* and *One Hundred Problems*.

The next lemma is stated here for its use in proving the theorem that follows it. It is interesting in its own right, since it characterizes completeness in terms of absolute summability in norm. We use f to denote an arbitrary element of a normed linear space mostly because we are now focusing our attention on spaces whose elements are functions. However, it should be noted that the lemma applies to all normed linear spaces.

Lemma 3.20. *A normed linear space* $(X, \| \cdot \|)$ *is complete if and only if* $\sum_{j=1}^{\infty} f_j$ *converges (in norm) whenever* $\sum_{j=1}^{\infty} \| f_j \|$ *converges.*

PROOF.   We start by assuming that X is complete, and consider a sequence $\{f_j\}_{j=1}^{\infty}$ in X such that $\sum_{j=1}^{\infty} \| f_j \|$ converges. Let $\epsilon > 0$. Since $\sum_{j=1}^{\infty} \| f_j \|$ converges, there exists N such that

$$\sum_{j=N}^{\infty} \| f_j \| < \epsilon.$$

Let s_n denote the nth partial sum of the series $\sum_{j=1}^{\infty} f_j$; that is $s_n = \sum_{j=1}^{n} f_j$. For $n \geq m \geq N$,

$$\| s_n - s_m \| = \left\| \sum_{j=m+1}^{n} f_j \right\| \leq \sum_{j=m+1}^{n} \| f_j \| \leq \sum_{j=m+1}^{\infty} \| f_j \| < \epsilon.$$

Since X is complete, $\{s_n\}_{n=1}^{\infty}$ converges.

To show the other direction we consider a Cauchy sequence $\{f_j\}_{j=1}^{\infty}$ in X. For each k there exists j_k such that

$$\| f_i - f_j \| < \frac{1}{2^k}, \quad i, j \geq j_k.$$

We may assume that $j_{k+1} > j_k$. This implies that $\{f_{j_k}\}_{k=1}^{\infty}$ is a subsequence of $\{f_j\}_{j=1}^{\infty}$. Set

$$g_1 = f_{j_1}, \text{ and } g_k = f_{j_k} - f_{j_{k-1}} \text{ for } k \geq 2.$$

Observe that

$$\sum_{k=1}^{l} \| g_k \| = \| g_1 \| + \sum_{k=2}^{l} \| f_{j_k} - f_{j_{k-1}} \| < \| g_1 \| + \sum_{k=2}^{l} \frac{1}{2^{k-1}} \leq \| g_1 \| + 1.$$

Therefore $\{\sum_{k=1}^{l} \| g_k \|\}_{l=1}^{\infty}$ is a bounded, increasing sequence which thus converges. By hypothesis, $\sum_{k=1}^{\infty} g_k$ converges. Since $\sum_{k=1}^{n} g_k = f_{j_n}$, it follows that $\{f_{j_n}\}_{n=1}^{\infty}$ converges in X. Let $f \in X$ denote the limit of the subsequence $\{f_{j_n}\}_{n=1}^{\infty}$ of $\{f_j\}_{j=1}^{\infty}$. We will be done when we show that $\{f_j\}_{j=1}^{\infty}$ also converges to f. Let $\epsilon > 0$. Since $\{f_j\}_{j=1}^{\infty}$ is Cauchy, there exists N such that

$$\| f_i - f_j \| < \frac{\epsilon}{2}, \quad i, j \geq N.$$

Also, there is a K such that

$$\|f_{j_k} - f\| < \frac{\epsilon}{2}, \quad k \geq K.$$

Choose k such that $k \geq K$ and $j_k \geq N$. By the triangle inequality we have

$$\|f_j - f\| \leq \|f_j - f_{j_k}\| + \|f_{j_k} - f\| < \epsilon, \quad j \geq N,$$

completing the proof. □

We now present one of F. Riesz's most important results.

Theorem 3.21. *For* $1 \leq p \leq \infty$, *$L^p(\mu)$ is complete.*

PROOF. We first do the case $p = \infty$; this is the easiest part of the proof. Let $\{f_j\}_{j=1}^\infty$ be an arbitrary Cauchy sequence in $L^\infty(\mu)$. By Exercise 3.6.7 there exist measure zero sets $A_{m,n}$ and B_j, $j, m, n = 1, 2, \ldots$, such that

$$|f_n(x) - f_m(x)| \leq \|f_n - f_m\|_\infty$$

for all $x \notin A_{m,n}$, and

$$|f_j(x)| \leq \|f\|_\infty$$

for all $x \notin B_j$. Define A to be the union of these sets for $j, m, n = 1, 2, \ldots$. Then A has measure zero (Exercise 3.2.6). Define

$$f(x) = \begin{cases} 0 & \text{if} \quad x \in A, \\ \lim_{j \to \infty} f_j(x) & \text{if} \quad x \notin A. \end{cases}$$

Then f is measurable. Also, for each $x \notin A$ there exists a positive integer N_x such that

$$|f_n(x) - f(x)| < 1, \quad n \geq N_x.$$

In particular,

$$|f_{N_x}(x) - f(x)| < 1, \quad x \notin A.$$

From this it follows that

$$|f(x)| < 1 + |f_{N_x}(x)| \leq 1 + \|f_{N_x}\|_\infty, \quad x \notin A,$$

and hence that $f \in L^\infty$. Since a Cauchy sequence is bounded, there exists $M > 0$ such that $\|f_j\|_\infty \leq M$ for every $j = 1, 2, \ldots$. In particular, $\|f_{N_x}\|_\infty \leq M$. The last inequality now shows that $f \in L^\infty(\mu)$.

Now we want to show that our given Cauchy sequence actually converges to this element f (in $L^\infty(\mu)$; we already know that it converges pointwise almost everywhere, but this is a weaker assertion than we need). To this end, let $\epsilon > 0$. There exists a positive integer K such that $n, m \geq K$ imply

$$\|f_n - f_m\|_\infty < \epsilon.$$

Then

$$|f_n(x) - f_m(x)| < \epsilon$$

almost everywhere, and so

$$\lim_{m \to \infty} |f_n(x) - f_m(x)| \le \epsilon,$$

for $n \ge K$ and $x \notin A$. This shows that

$$\|f - f_n\|_\infty \le \epsilon,$$

for $n \ge K$, as desired.

We next tackle the other cases, for $1 \le p < \infty$. The proof is not trivial, and the artillery required is somewhat substantial; we will use the lemma preceding this theorem, Fatou's lemma, Lebesgue's dominated convergence theorem, and Minkowski's inequality. Consider a Cauchy sequence $\{f_k\}_{k=1}^\infty$ in $L^p(\mu)$, $1 \le p < \infty$, such that

$$\sum_{k=1}^\infty \|f_k\|_p = M < \infty.$$

By Lemma 3.20, it suffices to show that $\sum f_k$ converges (in norm), that is, that there exists a function $s \in L^p$ such that

$$\left\| \left(\sum_{k=1}^n f_k \right) - s \right\|_p \to 0, \quad \text{as } n \to \infty.$$

We work on determining this s. Define, for each positive integer n,

$$g_n(x) = \sum_{k=1}^n |f_k(x)|.$$

Minkowski's inequality implies that

$$\|g_n\|_p \le \sum_{k=1}^n \|f_k\|_p \le M.$$

Therefore,

$$\int_X (g_n)^p d\mu \le M^p.$$

For each $x \in X$, $\{g_n(x)\}_{n=1}^\infty$ is an increasing sequence of numbers in $\mathbb{R} \cup \{\infty\}$, and so there exists a number $g(x) \in \mathbb{R} \cup \{\infty\}$ to which the sequence $\{g_n(x)\}_{n=1}^\infty$ converges. The function g on X defined in this way is measurable, and Fatou's lemma asserts that

$$\int_X g^p d\mu \le \liminf_{n \to \infty} \left(\int_X (g_n)^p d\mu \right) \le M^p.$$

In particular, this shows that $g(x) < \infty$ almost everywhere. For each x such that $g(x)$ is finite, the series $\sum_{k=1}^\infty f_k(x)$ is an absolutely convergent series. Let

$$s(x) = \begin{cases} 0 & \text{if} & g(x) \text{ is infinite,} \\ \sum_{k=1}^\infty f_k(x) & \text{if} & g(x) \text{ is finite.} \end{cases}$$

This function is equal, almost everywhere, to the limit of the partial sums $s_n(x) = \sum_{k=1}^{n} f_k(x)$, and hence is itself measurable. Since

$$|s_n(x)| \leq g(x),$$

we have that

$$|s(x)| \leq g(x).$$

Thus, $s \in L^p$ and

$$|s_n(x) - s(x)|^p \leq \left(|s_n(x)| + |s(x)|\right)^p \leq 2^p \left(g(x)\right)^p.$$

We can now apply Lebesgue's dominated convergence theorem to get

$$\lim_{n \to \infty} \left(\int_X (s_n - s)^p d\mu \right) = 0.$$

In other words,

$$\lim_{n \to \infty} \left(\|s_n - s\|_p \right)^p = 0,$$

and hence

$$\lim_{n \to \infty} \|s_n - s\|_p = 0,$$

which is precisely what we wanted to prove. $\square$

The most important L^p-spaces for us will be $L^p(X, \mu)$ where μ is either Lebesgue measure on some (not necessarily proper) subset of $\mathbb{R}^n$, or μ is counting measure on $X = \mathbb{N}$. In the case of counting measure, the L^p-space is denoted by $\ell^p(\mathbb{N})$, or ℓ^p (read "little ell p"), and *is* (the reader should come to grips with this assertion) the space of all sequences $\{x_n\}_{n=1}^{\infty}$ satisfying

$$\sum_{n=1}^{\infty} |x_n|^p < \infty,$$

with norm given by

$$\|\{x_n\}_{n=1}^{\infty}\|_p = \left(\sum_{n=1}^{\infty} |x_n|^p \right)^{\frac{1}{p}}.$$

Note that ℓ^∞ is the set of all bounded sequences, with

$$\|\{x_n\}_{n=1}^{\infty}\|_\infty = \sup\{|x_n| \,\big|\, n = 1, 2, \ldots\}.$$

(Recall the material of Section 1.2.)

Theorem 3.21 shows that all L^p-spaces, $1 \leq p \leq \infty$, are complete. The cases $1 \leq p < \infty$ are deeper than the case $p = \infty$, and it is harder to supply a proof that applies for all measures. There are, however, some specific measures for which there are easier proofs. It is instructive to see some of these as well, and for this reason an alternative proof of the completeness of ℓ^p, $1 \leq p < \infty$, is now given.

We consider a Cauchy sequence $\{x_n\}_{n=1}^{\infty}$ (a sequence of sequences!) in ℓ^p, with the sequence x_n given by

$$x_n = \left(a_1^{(n)}, a_2^{(n)}, \ldots\right).$$

For a fixed k, observe that

$$\left|a_k^{(n)} - a_k^{(m)}\right| \le \left(\sum_{i=1}^{\infty} \left|a_i^{(n)} - a_i^{(m)}\right|^p\right)^{\frac{1}{p}}.$$

This shows that for each fixed k, the sequence $\{a_k^{(n)}\}_{n=1}^{\infty}$ is a Cauchy sequence of real numbers. Therefore, $\{a_k^{(n)}\}_{n=1}^{\infty}$ converges; let

$$a_k = \lim_{n \to \infty} a_k^{(n)}.$$

We now show two things:

(i) $a = \{a_k\}_{k=1}^{\infty}$ is in ℓ^p;
(ii) $\lim_{n \to \infty} \|x_n - a\|_p = 0.$

Exercise 2.3.1 shows that there exists M such that

$$\|x_n\|_p \le M,$$

for all $n = 1, 2, \ldots$. For any k, we thus have,

$$\left(\sum_{i=1}^{k} |a_i^{(n)}|^p\right)^{\frac{1}{p}} \le \|x_n\|_p \le M.$$

Letting $n \to \infty$ yields

$$\left(\sum_{i=1}^{k} |a_i|^p\right)^{\frac{1}{p}} \le M.$$

Since k is arbitrary, this shows that $a = \{a_k\}_{k=1}^{\infty}$ is in ℓ^p, and also that

$$\|a\|_p \le M.$$

To show (ii), let $\epsilon > 0$. Then there exists a positive integer N such that

$$\|x_n - x_m\|_p < \epsilon, \qquad n, m \ge N.$$

For any k, we thus have,

$$\left(\sum_{i=1}^{k} |a_i^{(n)} - a_i^{(m)}|^p\right)^{\frac{1}{p}} \le \|x_n - x_m\|_p < \epsilon, \qquad n, m \ge N.$$

If we now keep both n and k fixed, and let $m \to \infty$, we get

$$\left(\sum_{i=1}^{k} |a_i^{(n)} - a_i|^p\right)^{\frac{1}{p}} < \epsilon, \qquad n \ge N.$$

Since k is arbitrary, this shows that $\|x_n - a\|_p \le \epsilon$ for $n \ge N$, which is equivalent to (ii).

Note that the outline for the proof in the special case of counting measure is exactly the same as for the general measure proof given in Theorem 3.21: Take a specially designed function (sequence), which is a "pointwise" limit of sorts of the given Cauchy sequence, and then

(i) prove that this specially designed function (sequence) is in fact an element of the space;
(ii) prove that the convergence is in fact in norm (and not just "pointwise").

The difficulty encountered in the proof in the general case comes in having to prove these two properties for *arbitrary* measures.

We end this chapter with a big theorem, proved (for Lebesgue measure on Euclidean space) by F. Riesz in the 1910 paper [105]. Recall the definition of a simple function. A simple function is called a *step function* if each of the sets E_k has finite measure.

Theorem 3.22. *The step functions are dense in $L^p(\mu)$, for each $1 \le p < \infty$.*

PROOF. Let $0 \le f \in L^p$. By Theorem 3.10 we can construct a sequence of simple functions, $\{s_n\}_{n=1}^\infty$, such that

$$0 \le s_1 \le s_2 \le \cdots \le f, \quad \lim_{n \to \infty} s_n(x) = f(x) \text{ almost everywhere.}$$

Each of these simple functions is, in fact, a step function. Furthermore,

$$(f-s_1)^p \ge (f-s_2)^p \ge \cdots \ge 0, \quad \text{and} \quad \lim_{n \to \infty} ((f-s_n)(x))^p = 0 \text{ almost everywhere.}$$

Lebesgue's dominated convergence theorem now tells us that

$$\|f - s_n\|_p = \lim_{n \to \infty} \left(\int_X |f - s_n|^p \right)^{\frac{1}{p}} = 0.$$

Since every element of L^p can be written as the difference of two nonnegative functions in L^p, the proof is done. □

We end this section by remarking that L^2 is a Hilbert space. What remains to be seen in this is that the norm comes from an inner product. This is easily seen, by defining

$$\langle f, g \rangle = \int_X f \bar{g} d\mu$$

for complex-valued functions. The Hilbert space L^2 will be discussed in great detail in the next chapter. One can also show that the norm on L^p, for $p \ne 2$, does not come from an inner product. A proof of this is outlined in Exercise 3.6.9.

Frigyes Riesz was born in Gyor, Austria–
Hungary (in what is now Hungary)
on January 22, 1880 (Figure 3.2). His
father was a physician. His younger
brother Marcel was also a distinguished
mathematician.

Frigyes Riesz studied in Budapest
and then went to Göttingen and Zurich
before returning to Budapest, where
he received his doctorate in 1902. His
dissertation built on ideas of Fréchet, and
made connections between Lebesgue's
work on measure-theoretic notions and
the work of Hilbert and his student
Erhard Schmidt (1876–1959; Russia, now
Estonia) on integral equations. Hilbert and
Schmidt had been working with integral
equations in which the functions were
assumed continuous. Riesz, in this context,
introduced the Lebesgue square integrable
(L^2-) functions. He was also interested in
knowing which sequences of real numbers
could arise as the Fourier coefficients
of some function. He answered this
question, as did Fischer, and the result is
known as the Riesz–Fischer theorem (see
Section 4.2).

Over the next few years, and in an
attempt to generalize the Riesz–Fischer
theorem, Riesz introduced the L^p-spaces
for $p > 1$, and the general theory of
normed linear spaces. One of the most
important results about the L^p-spaces is
his Riesz representation theorem. This
theorem completely describes all the continuous linear functionals (see Section 6.3)
from L^p to $\mathbb{C}$. Riesz is often considered to
be the "father" of abstract operator theory.
Hilbert's eigenvalue problem for integral
equations was dealt with quite effectively
by Riesz in this more abstract setting. Riesz
was able to obtain many results about the
spectra of the integral operators

FIGURE 3.2. Frigyes Riesz.

associated with the integral equations of
Hilbert.

As alluded to in the preceding paragraph, Riesz also introduced the notion of a
norm, but this idea did not come to fruition
until Banach wrote down his axioms for a
normed linear space in [10].

Frigyes Riesz made many important
contributions to functional analysis, as
well as to the mathematics profession as
a whole. His ideas show great originality
of thought, and aesthetic judgment in
mathematical taste. He is one of the
founders of the general theory of normed
linear spaces and the operators acting on
them. His theory of compact operators,
which generalizes work of Fredholm,
set the stage for future work on classes
of operators. While he did so much on
this abstract theory, Riesz was originally
motivated by very concrete problems,
and often returned to them in his work.
Most of Riesz's work on operator theory in
general, and spectral theory in particular,
lies beyond the scope of this book. For
a detailed historical account of Riesz's
contributions, see [34] or [94].

Riesz was able to communicate about mathematics superbly. He wrote several books and many articles, and served as editor of the journal *Acta Scientiarum Mathematicarum*. His book [107] is a classic that continues to serve as an excellent introduction to the subject.

Riesz died on February 28, 1956, in Budapest.

Exercises for Chapter 3

Section 3.1

3.1.1 Write out the details of the proof, using diagonalization, that $\mathcal{B}$ is uncountable.

3.1.2 Prove that $\mathcal{B}_T$ is countable, and use this result to give an alternative proof (as outlined in the text) that $\mathcal{B}$ is uncountable.

3.1.3 In this exercise you are asked to determine the set B_E for given events E.

(a) Determine the set B_E if E is the event that in the first three tosses, exactly two heads are seen. What is the probability of this event occurring? Does your answer to the last question agree with what you think the Lebesgue measure of B_E should be?

(b) Determine the set B_E if E is the event that in the first n tosses, exactly k heads are seen. What is the probability of this event occurring? Does your answer to the last question agree with what you think the Lebesgue measure of B_E should be? Explain.

3.1.4 Prove that $(0, 1] \setminus S$ is uncountable, where S is the set referred to in the strong law of large numbers. Hint: Consider the map from $(0, 1]$ to itself that maps the binary expression $\omega = .a_1a_2\ldots$ to $.a_111a_211a_311a_4\ldots$. Prove that this map is one-to-one and its image is contained in $(0, 1] \setminus S$.

Section 3.2

3.2.1 Let $\mathcal{R}$ be a σ-ring. Prove that $\bigcap_{n=1}^{\infty} A_n \in \mathcal{R}$ whenever $A_n \in \mathcal{R}$, $n = 1, 2, \ldots$. Hint: Verify, and use, that $\bigcap_{n=1}^{\infty} A_n = A_1 \setminus \bigcup_{n=1}^{\infty}(A_1 \setminus A_n)$.

3.2.2 Assume that μ is a nonnegative, additive function defined on a ring $\mathcal{R}$.

(a) Prove that μ is monotone; that is, show that $\mu(A) \leq \mu(B)$ whenever $A, B \in \mathcal{R}$ and $A \subseteq B$.

(b) Prove that μ is *finitely subadditive*; that is, show that

$$m\left(\bigcup_{k=1}^{n} A_k\right) \leq \sum_{k=1}^{n} m(A_k)$$

whenever $A_1, A_2, \ldots \in \mathcal{R}$.

3.2.3 Assume that μ is a countably additive function defined on a ring $\mathcal{R}$, that $A_n \in \mathcal{R}$, $A \in \mathcal{R}$, that

$$A_1 \subseteq A_2 \subseteq \cdots,$$

and that

$$A = \bigcup_{n=1}^{\infty} A_n.$$

Prove that

$$\lim_{n \to \infty} \mu(A_n) = \mu(A).$$

(Hint: Put $B_1 = A_1$, and $B_n = A_n \setminus A_{n-1}$ for $n = 2, 3, \ldots$.)

3.2.4 Prove that $\mathcal{E}$ is a ring, but is not a σ-ring.

3.2.5 Prove Lemma 3.1. (Hint: Consider the case that A is an interval first, and then consider finite unions of disjoint intervals.)

3.2.6 Prove that m^* is countably subadditive.

3.2.7 This exercise is about the symmetric difference and distance functions S and D defined on 2^X.

   (a) Let $A = [0, 4] \times (1, 10]$ and $B = (0, 1] \times [0, 2]$ in $\mathbb{R}^2$. Draw a picture of the set $S(A, B)$, and compute $D(A, B)$.

   (b) Consider arbitrary subsets A and B of an arbitrary set X. Prove that $D(A, B) = D(B, A)$.

   (c) Consider arbitrary subsets A and B of an arbitrary set X. Does $D(A, B) = 0$ necessarily imply that $A = B$? Either prove that it does, or give a counterexample to show that it does not.

   (d) Consider arbitrary subsets A, B, C of an arbitrary set X. Prove that

$$S(A, C) \subseteq S(A, B) \cup S(B, C),$$

   and deduce that

$$D(A, C) \le D(A, B) + D(B, C).$$

   (e) Consider arbitrary subsets A_1, A_2, B_1, B_2 of an arbitrary set X. Prove

$$S(A_1 \cup A_2, B_1 \cup B_2) \subseteq S(A_1, B_1) \cup S(A_2, B_2),$$

   and deduce that

$$D(A_1 \cup A_2, B_1 \cup B_2) \le D(A_1, B_1) + D(A_2, B_2).$$

3.2.8 If you have studied some abstract algebra, you may know a different use of the term "ring" (the definition is given, incidentally, at the beginning of Section 6.6). In this exercise, the term "ring" refers to the algebraic notion. Prove that $2^{(\mathbb{R}^n)}$ becomes a commutative ring with "multiplication" of two sets taken to be their intersection, and with "addition" of two sets taken to be their symmetric difference.

3.2.9 **(a)** Prove that all open subsets of $\mathbb{R}^n$ are in $\mathcal{M}$.

(b) Prove that all closed subsets of $\mathbb{R}^n$ are in $\mathcal{M}$.

(c) Prove that all countable unions and intersections of open and closed subsets of $\mathbb{R}^n$ are in $\mathcal{M}$.

3.2.10 Prove that the Cantor set is in $\mathcal{M}$ and that it has Lebesgue measure zero.

Section 3.3

3.3.1 Prove that every continuous function is measurable.

3.3.2 Give an example of a function f such that f is not measurable but $|f|$ is measurable.

3.3.3 Which characteristic functions are (Lebesgue) integrable on $\mathbb{R}$? Is the characteristic function of the rational numbers integrable on the unit interval? If so, what is value of this integral?

3.3.4 Prove Theorem 3.8. (Hint: One approach is to notice that

$$\{x \mid |f(x)| > a\} = \{x \mid f(x) > a\} \cup \{x \mid f(x) < -a\}.)$$

3.3.5 Fill in the details of the proof of Theorem 3.9.

3.3.6 Complete the proof of Theorem 3.10.

3.3.7 Prove that $f \in \mathcal{L}(\mathbb{R}^n)$ if and only if $\int_{\mathbb{R}^n} |f| dm < \infty$.

3.3.8 Supply proofs for the missing parts of Theorem 3.11.

(a) Prove the first part of part (a) of Theorem 3.11.

(b) Prove the second part of part (a) of Theorem 3.11, for simple functions. (The general case appears after the proof of the monotone convergence theorem.)

(c) Prove part (b) of Theorem 3.11.

(d) Prove part (c) of Theorem 3.11.

(e) Prove part (d) of Theorem 3.11.

3.3.9 You have read about the phrase "almost everywhere" in the text. In particular, we say that two measurable functions are "equal almost everywhere" if the set of points where they differ has measure zero.

(a) Prove that this relation is an equivalence relation on the set of integrable functions.

(b) Prove that f and g are equal almost everywhere if and only if

$$\int_E f dm = \int_E g dm$$

for every measurable set E.

(c) Prove that $f = 0$ almost everywhere if $\int_E f dm = 0$ for every $E \in \mathcal{M}$.

Section 3.4

3.4.1 Give an example to show that strict inequality can hold in Fatou's lemma.

3.4.2 Give an example to show that without the existence of the function g in the dominated convergence theorem, the conclusion may fail.

Section 3.6

3.6.1 Prove that the triples $(X, \mathcal{R}, \mu)$ given in Examples 4 and 6 at the beginning of Section 6 are, in fact, measure spaces. For the measure in Example 3, show that

$$\mu\left(\bigcup_{n=1}^{\infty} [a_n, b_n]\right) = \sum_{n=1}^{\infty} \mu([a_n, b_n])$$

whenever $n \neq m$ implies $[a_n, b_n] \cap [a_m, b_m] = \emptyset$.

3.6.2 For $0 < p < 1$, we can define the L^p-space and $\|\cdot\|_p$ in the same way that we did for $1 \leq p < \infty$. Prove, by giving a suitable example, that $\|\cdot\|_p$ does not satisfy the triangle inequality, and hence is not a norm.

3.6.3 Prove, as outlined in the text, that

$$xy \leq \frac{x^p}{p} + \frac{y^q}{q},$$

for $x, y \geq 0$ and Hölder conjugates p and q with $1 < p < \infty$.

3.6.4 Prove that

$$\| |f + g|^{p-1} \|_q = \left(\|f + g\|_{(p-1)q}\right)^{p-1} = \left(\|f + g\|_p\right)^{p-1},$$

for $f, g \in L^p(\mu)$ and $1 < p < \infty$. (This equality is used in the proof of Theorem 3.19.)

3.6.5 Prove that L^∞, with norm $\|\cdot\|_\infty$, is a normed linear space.

3.6.6 In this exercise you will investigate relations between the various L^p-spaces.

(a) Let $1 \leq p < q \leq \infty$. Consider Lebesgue measure on $\mathbb{R}^n$. Construct examples to show that neither $L^p \subseteq L^q$ nor $L^q \subseteq L^p$ holds.

(b) Next, suppose that $1 \leq p < r < q < \infty$. Show that $L^p \cap L^q \subseteq L^r$. (This is for any measure space $(X, \mathcal{R}, \mu)$.)

(c) Now assume that X is a *finite* measure space, i.e., that X is measurable and that $\mu(X) < \infty$. Prove that $L^q \subseteq L^p$ for $1 \leq p < q < \infty$, and give an example to show that this is a proper inclusion. Now prove that

$$\|f\|_1 \leq \|f\|_p \leq \|f\|_q \leq \|f\|_\infty, \quad 1 \leq p < q < \infty,$$

whenever μ is a probability measure.

(d) Prove that $\ell^p \subseteq \ell^q$ for $1 \leq p < q \leq \infty$, and give an example to show that this is a proper inclusion. Now prove that

$$\|f\|_\infty \leq \|f\|_q \leq \|f\|_p \leq \|f\|_1, \quad 1 \leq p < q < \infty.$$

3.6.7 Show that for each $f \in L^\infty$, $|f(x)| \leq \|f\|_\infty$ almost everywhere.

3.6.8 Let I be an interval in $\mathbb{R}$.

(a) Assume that $1 \leq p < \infty$. Prove that $C(I)$ is dense in $L^p(m)$ whenever I is closed and bounded.

(b) Now drop the assumption that I is closed and bounded. We certainly cannot expect $C(I)$ to be dense in $L^p(m)$, since it is not even contained in it. We define the "continuous functions with compact support" on $X \subseteq \mathbb{R}$ to be

$$C_c(X) = \{f \mid f \in C(X) \text{ and } \overline{\{x \in X \mid f(x) \neq 0\}} \text{ is compact}\}.$$

The set $\overline{\{x \in X \mid f(x) \neq 0\}}$ is often called the "support" of f. Again, assume that $1 \leq p < \infty$. Prove that $C_c(I)$ is dense in $L^p(m)$. (Note that this result subsumes the result of (a).)

(c) Prove that $C_c(I)$ is not dense in $L^\infty(m)$.

(The results of Exercise 8 give an alternative proof of Theorem 4.7, and give a way to define L^p that is independent of measure theory. See the paragraph following the proof of Theorem 4.7.)

3.6.9 The point of this exercise is to show that the norm on L^p, for $p \neq 2$, does not come from an inner product. We start by considering $L^p([-1,1])$, with Lebesgue measure. Set $f(x) = 1 + x$ and $g(x) = 1 - x$.

(a) Show that

$$\|f\|_p^p = \frac{2^{p+1}}{p+1} = \|g\|_p^p,$$
$$\|f + g\|_p^p = 2^{p+1},$$
$$\|f - g\|_p^p = \frac{2^{p+1}}{p+1}.$$

(b) Using part (a), show that the parallelogram equality asserts that

$$(p+1)^{\frac{2}{p}} = 3.$$

Verify that this equality holds for $p = 2$.

(c) Prove that the parallelogram equality does not hold for values $p \neq 2$. Hint: Show that the function $(p+1)^{\frac{2}{p}} - 3$ is a strictly decreasing function of $p \geq 1$ and thus takes on the value zero for at most one value of p.

(d) Modify the functions given in part (a) to prove the result for $L^p(I)$ where I is any interval, bounded or unbounded, in $\mathbb{R}$.

4

Fourier Analysis in Hilbert Space

In the last section of Chapter 3 we introduced the Lebesgue L^p-spaces for general measures and discussed their most basic properties. The most important L^p-space, by far, is L^2. Its importance is its role in applications, especially in Fourier analysis. The material of this chapter lies at the foundation of the branch of mathematics called harmonic analysis.

In this chapter we will see that L^2 is a Hilbert space (we already really have all the bits of information we need to see this) and that in some sense the L^2-spaces (with different μ's) are the *only* Hilbert spaces. We will come to see how the problem that Fourier examined, about decomposing functions as infinite sums of other — somehow more basic — functions, is a problem best phrased and understood in the language of abstract Hilbert spaces. One of the triumphs of functional analysis is to take a very concrete problem — in this case Fourier decomposition — view it in an abstract setting, and use theoretical tools to obtain powerful results that can be translated back to the concrete setting. Fourier's work certainly holds an important spot at the roots of functional analysis, and it motivated much early work in the development of the field.

Further Hilbert space theory appears in Section 5.4.

4.1 Orthonormal Sequences

During the second half of the eighteenth century and first decade of the nineteenth century, infinite sums of sines and cosines appeared as solutions to physical problems then being studied. Daniel Bernoulli (1700–1782; Netherlands)[1] suggested that these sums were solutions to the problem of modeling the vibrating string, and Joseph Fourier (1768–1830; France) proposed them as solutions to the problem of modeling heat flow. It is not really until the response to Fourier's work that we see other mathematicians coming to grips with the challenge that these infinite sums truly posed: to understand the fundamental notions of *convergence* and *continuity*. Over the decades following the appearance of Fourier's works on heat, the field of "real analysis" would be born in large part out of efforts to respond to the challenges that Fourier's work raised in pure mathematics. Many of the great mathematicians of the period — perhaps most notably Cauchy, Riemann, and Weierstrass — did their most important work in the development of this field. For an excellent historical account of these mathematical developments, see [25]. Fourier begins with an arbitrary function f on the interval from $-\pi$ to π and states that *if* we can write

$$f(x) = \frac{a_0}{2} + \sum_{k=1}^{\infty} a_k \cos(kx) + b_k \sin(kx),$$

then it must be the case that the coefficients a_k and b_k are given by the formulas

$$a_k = \frac{1}{\pi} \int_{-\pi}^{\pi} f(x) \cos(kx) dx, \quad k = 0, 1, 2, \ldots,$$

and

$$b_k = \frac{1}{\pi} \int_{-\pi}^{\pi} f(x) \sin(kx) dx, \quad k = 1, 2, \ldots.$$

The big question is this: *When* is this decomposition actually possible? Even if the integrals involved make sense, does the series converge? If it does converge, what *type* of convergence (pointwise, uniform, etc.) do we get? Even if the series converges in some sense, does it converge to f?

The immediate goal is to show you how these questions about Fourier series can be treated in the abstract setting of an inner product space.

Let us now take stock of what we already know by gathering our information about L^2. First, recall that $L^2 = L^2(\mu)$, for any abstract measure space $(X, \mathcal{R}, \mu)$, denotes the collection of all measurable functions $f : X \to \mathbb{C}$ such that the integral

$$\int_X |f|^2 d\mu$$

---

[1] Daniel Bernoulli is the nephew of James Bernoulli, who was mentioned at the beginning of Section 3.1. The Bernoulli family produced several distinguished mathematicians and physicists; at least twelve members of the family achieved distinction in at least one of these fields.

is finite. These functions are often called the "square integrable" functions on X. With norm

$$\|f\|_2 = \sqrt{\int_X |f|^2 d\mu},$$

this collection of functions becomes a Banach space. We can define an inner product on L^2 via

$$\langle f, g \rangle = \int_X f\bar{g} d\mu.$$

It is easily seen that this is an inner product, and that the norm does indeed come from this inner product. That is,

$$\|f\|_2 = \sqrt{\langle f, f \rangle} = \sqrt{\int_X |f|^2 d\mu}.$$

Theorem 3.21 shows that L^2 is a Hilbert space.

In the following definitions, the terminology should seem familiar from your experiences with $\mathbb{R}^n$.

Let $(V, \langle \cdot, \cdot \rangle)$ be an inner product space. We say that v and w in V are *orthogonal* if $\langle v, w \rangle = 0$. We say that v is *normalized* if $\|v\| = \sqrt{\langle v, v \rangle} = 1$. A sequence $\{v_k\}_{k=1}^\infty$ in V is an *orthonormal sequence* if $\langle v_k, v_j \rangle = \delta_{kj}$, $1 \le k, j < \infty$. The function δ_{kj} is defined to be 1 if $k = j$ and 0 if $k \ne j$.

In Exercise 4.1.1 you are asked to show that the *trigonometric system* (Figure 4.1)

$$\frac{1}{\sqrt{2\pi}}, \quad \frac{\cos(nx)}{\sqrt{\pi}}, \quad \frac{\sin(mx)}{\sqrt{\pi}}, \quad n, m = 1, 2, \ldots,$$

is an orthonormal sequence in the inner product space $L^2([-\pi, \pi], m)$. From this, you should find it plausible that the goal of Fourier analysis in its general setting is this: Given an orthonormal sequence $\{f_k\}_{k=1}^\infty$ in an inner product space V and an $f \in V$, find complex numbers c_k such that

$$f = \sum_{k=1}^\infty c_k f_k.$$

The convergence of this infinite sum is in the norm induced by the inner product. Further, it would be desirable to be able to do this *for all* $f \in V$. In general, this cannot be done. Notice that Fourier was asserting that when $\{f_k\}_{k=1}^\infty$ is the trigonometric system, the coefficients are of form $\langle f, f_k \rangle$ (an appropriate indexing of the trigonometric system has not yet been established) whenever his decomposition works.

Let $\{f_k\}_{k=1}^\infty$ be an orthonormal sequence in V. If it is the case that for each $f \in V$ we can find constants c_k (depending on f) such that

$$f = \sum_{k=1}^\infty c_k f_k,$$

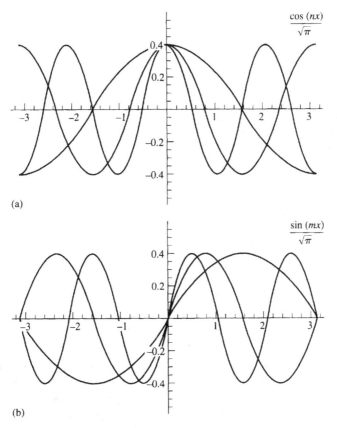

FIGURE 4.1. (a) the functions $\frac{\cos(nx)}{\sqrt{\pi}}$ for $n = 1, 2, 3$. (b) the functions $\frac{\sin(mx)}{\sqrt{\pi}}$ for $m = 1, 2, 3$.

then we say that the sequence $\{f_k\}_{k=1}^{\infty}$ is a *complete orthonormal sequence* in V.[2] A complete orthonormal sequence is sometimes called an *orthonormal basis* for V. The latter terminology can cause confusion since a complete orthonormal system is not a basis in the finite-dimensional sense discussed in Section 1.3.

The questions posed by Fourier's work are, to some degree, answered by the fact that the trigonometric system does indeed form a complete orthonormal sequence in L^2. This important result appears as Theorem 4.6.

The trigonometric system is certainly an important complete orthonormal sequence (for the Hilbert space $L^2([-\pi, \pi])$). But there are others, and we end this section with a brief description of a few of them ([43] is a good general reference for this topic). We can use the Gram–Schmidt process to construct an orthonormal sequence in any inner product space.

For our first example, the Hilbert space is $L^2([-1, 1])$. If one applies the Gram–Schmidt process to the functions $1, x, x^2, x^3, \ldots$, one obtains the complete

---

[2]Note that this is a new usage of the word "complete"; we now have at least two ways we will use this adjective: a *complete* metric space, a *complete* orthonormal system.

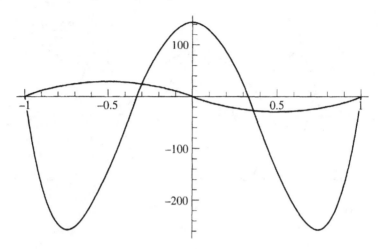

FIGURE 4.2. The Legendre polynomials, $n = 3, 4$.

orthonormal sequence of *Legendre polynomials* (Figure 4.2),

$$\sqrt{\frac{2n + 1}{2}} \frac{1}{2^n n!} \frac{d^n}{dx^n} (x^2 - 1)^n, \qquad n = 1, 2, \dots .$$

These polynomials are named for Adrien-Marie Legendre (1752–1833; France).

Next, consider the Hilbert space $L^2((0, \infty))$. If one applies the Gram–Schmidt process to the the functions $x^n e^{-x}$, $n = 0, 1, \dots$, one obtains the complete orthonormal sequence of *Laguerre functions*. These appear in quantum mechanics in the analysis of the hydrogen atom. This family is named for Edmond Laguerre (1834–1886; France).

For our third example, the Hilbert space is $L^2(\mathbb{R})$. If one applies the Gram–Schmidt process to the the functions $x^n e^{\frac{-x^2}{2}}$, $n = 0, 1, \dots$, one obtains the complete orthonormal sequence of *Hermite functions*. These also appear in quantum mechanics. This family is named for Charles Hermite (1822–1901; France).

The Legendre, Laguerre, and Hermite functions all show up as eigenfunctions of certain linear operators (linear operators are the subject of the next chapter) related to the Sturm–Liouville problem in differential equations.

The final family we discuss is the complete orthonormal sequence of *Haar functions*. The Hilbert space is $L^2([0, 1])$. This example is fundamentally different from the previous examples in that the functions in this family are not continuous, and they are not connected with differential equations. Haar functions appear in the study of "wavelets." Wavelet theory and its applications experienced explosive development in the 1980s. There are several good books, at varying levels, on the subject. A "brief" investigation of wavelets, their properties and uses, makes a good student project ([28] gives an excellent overview and introduction to wavelets). Wavelet series are used in signal and imaging processing and, in some contexts,

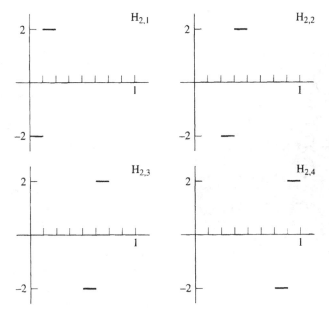

FIGURE 4.3. The Haar functions, $H_{2,k}(x)$.

are replacing the classical Fourier series. We define

$$H_{0,0}(x) = 1, \qquad H_{n,k}(x) = \begin{cases} -2^{\frac{n}{2}} & \text{if } \frac{k-1}{2^n} \leq x < \frac{k-\frac{1}{2}}{2^n}, \\ 2^{\frac{n}{2}} & \text{if } \frac{k-\frac{1}{2}}{2^n} \leq x < \frac{k}{2^n}, \\ 0 & \text{otherwise}, \end{cases}$$

for $n \geq 1$, $1 \leq k \leq 2^n$ (Figure 4.3). This family is named for Alfréd Haar (1885–1933; Hungary).

Jean Baptiste Joseph Fourier was born March 21, 1768, in Auxerre, France (Figure 4.4). His father had been a tailor, but both of his parents were dead by the time Fourier was ten. There seems to be some disagreement among authors as to exactly how many siblings Fourier had, but by all accounts he had many. According to [51], he was the nineteenth (and not the last) child in the family.

Fourier was distinguished in two fields: mathematics and Egyptology. He began both careers when he attended a military school run by the Benedictines. He showed great talent in many areas by the age of fourteen. He wanted to join the military, for some reason was rejected, and instead entered a Benedictine abbey to train for the priesthood. While there, he was able to work on mathematics and submitted his first paper in 1789. He never took his vows and returned to his school, teaching math, history, philosophy, and rhetoric. This was the time of the French Revolution, and Fourier became quite involved in revolutionary politics. In 1794 he was imprisoned and sentenced to be guillotined.

FIGURE 4.4. Joseph Fourier.

In 1795 the École Normale in Paris opened to train teachers in an effort to rehabilitate the system of higher education in France. The students were chosen and financed by the republic. Fourier was chosen to attend, and while there, he came into contact with very good professors: Lagrange, Laplace, and Gaspard Monge (1746–18181; France). Unfortunately, the school closed after a few months. At this time, Fourier went to teach at the École Polytechnique, which was designed as a military academy to train the military elite. During this period he was, for a second time, arrested and subsequently freed.

Over the next few years, Fourier taught (mathematics with military applications) and worked on mathematical research (mostly having to do with polynomials: extending Descartes's rule of signs, approximating values of real roots, detecting existence of complex roots).

In 1798 Fourier was recommended, by Monge and the chemist Claude Louis Berthollet (1748–1822; France), to be Napoleon Bonaparte's scientific advisor on his expedition to Egypt. Very soon after their arrival in Egypt, the Institut d'Égypte opened in Cairo, and Fourier was appointed *secrétaire perpetuel*. He had many duties in this post, including investigating ancient monuments and irrigation projects, but he managed to find time to continue mathematical research.

Napoleon left for France in 1799. Fourier followed in 1801 and was appointed by Napoleon to a government position in Grenoble. He held this post from 1802 until 1814. During this period, he devoted much time to the writing of a massive work entitled *Description de l'Égypte*. This work was written by the team that Napoleon brought with him to Egypt and is very important in the birth of the modern field of Egyptology; it gave the most comprehensive account, to date, of ancient and contemporary Egypt. To put this accomplishment in perspective, the Rosetta Stone was discovered by this team, and it was in 1822 that hieroglyphics were fully deciphered.

It was also during his time in Grenoble that Fourier did his work on heat diffusion. This work, done primarily during the period 1804–1807, culminated in a monograph that was submitted to the Institut de France in Paris at the end of 1807. This paper caused a great deal of controversy. One complaint was from Lagrange and had to do with the convergence of his "Fourier series." Lagrange's skepticism was on target and, indeed, led to the rise of a new field of mathematics: "real analysis" (see [25]). The controversy caused Fourier to revise the paper and resubmit it in 1811. Eventually, his *Théorie analytique de la chaleur* was published in 1822. This work is Fourier's greatest contribution and certainly remains one of the masterpieces of mathematical physics. It is important not only for the physical explanations that it gives, but also for the mathematical techniques developed in the course of his attempting to explain the physics of heat flow. For example, he developed techniques to find solutions for many differential equations that, up until that point, had not been worked out.

In the last fifteen years of his life, Fourier continued to work on mathematics and on topics related to his work in Egypt. However, his most substantial contributions had already been made, and much of his mathematical work during his later years focused on consequences of his earlier work. One of his other important mathematical projects during this time was on problems that can now be viewed as precursors to the field of linear programming. He also did editorial work and wrote several biographies of mathematicians during this period.

Fourier died on May 16, 1830, after being in a state of deteriorating health for several years.

4.2   Bessel's Inequality, Parseval's Theorem, and the Riesz–Fischer Theorem

Let $(V, \langle \cdot, \cdot \rangle)$ be an inner product space, and $\{f_k\}_{k=1}^{\infty}$ a specified orthonormal sequence in V. Suppose we have an $f \in V$ that we can decompose as

$$f = \sum_{k=1}^{\infty} c_k f_k.$$

What, then, are the c_k's to be? We turn to the very simple case of $\mathbb{R}^3$, with its usual inner product, for inspiration. We take as our orthonormal family the three Euclidean basis vectors $e_1 = (1, 0, 0)$, $e_2 = (0, 1, 0)$, and $e_3 = (0, 0, 1)$. Then every vector in $\mathbb{R}^3$ can be written in form

$$\sum_{k=1}^{3} c_k e_k.$$

In this case we know that $c_1 = \langle v, e_1 \rangle$, $c_2 = \langle v, e_2 \rangle$, and $c_3 = \langle v, e_3 \rangle$. This example illustrates the next theorem.

Theorem 4.1. *Suppose that $f = \sum_{k=1}^{\infty} c_k f_k$ for an orthonormal sequence $\{f_k\}_{k=1}^{\infty}$ in an inner product space V. Then $c_k = \langle f, f_k \rangle$ for each k.*

Before we prove this result, notice that this is, in fact, consistent with Fourier's assertion about the trigonometric system.

PROOF.   Let $s_n = \sum_{k=1}^{n} c_k f_k$. Our hypothesis is thus that

$$\lim_{n \to \infty} \|s_n - f\| = 0.$$

Fix an m and let $n \geq m$. Then

$$\lim_{n \to \infty} \langle s_n, f_m \rangle = \langle f, f_m \rangle. \tag{4.1}$$

This is because

$$\langle s_n, f_m \rangle - \langle f, f_m \rangle = \langle s_n - f, f_m \rangle,$$

and

$$|\langle s_n - f, f_m \rangle| \le \| s_n - f \| \cdot \| f_m \| = \| s_n - f \|.$$

Therefore,

$$\langle s_n, f_m \rangle = \sum_{k=1}^{\infty} \langle c_k f_k, f_m \rangle = \sum_{k=1}^{\infty} c_k \langle f_k, f_m \rangle = \sum_{k=1}^{\infty} c_k \delta_{km} = c_m. \qquad (4.2)$$

Combining (4.1) and (4.2) gives the desired result.   □

Let $\{f_k\}_{k=1}^{\infty}$ be an orthonormal sequence in V, and let $f \in V$. We call $\sum_{k=1}^{\infty} \langle f, f_k \rangle f_k$ the *Fourier series* of f with respect to $\{f_k\}_{k=1}^{\infty}$, and $\langle f, f_k \rangle$ the *Fourier coefficients* of f with respect to $\{f_k\}_{k=1}^{\infty}$. These objects are defined without any assumptions or knowledge about convergence of the series.

The next theorem tells us something about the size of these coefficients.

Theorem 4.2 (Bessel's Inequality[3]).   *Suppose that $\{f_k\}_{k=1}^{\infty}$ is an orthonormal sequence in an inner product space V. For every $f \in V$, the series (of nonnegative real numbers) $\sum_{k=1}^{\infty} |\langle f, f_k \rangle|^2$ converges and*

$$\sum_{k=1}^{\infty} |\langle f, f_k \rangle|^2 \le \| f \|^2.$$

PROOF.   Consider the partial sum s_n of the Fourier series for f. Then

$$S_n = \sum_{k=1}^{n} C_k f_k$$
$$C_k = \langle f_1, f_k \rangle$$

$$\langle f - s_n, f_k \rangle = \langle f, f_k \rangle - \langle s_n, f_k \rangle$$

$$= \langle f, f_k \rangle - \left\langle \sum_{j=1}^{n} \langle f, f_j \rangle f_j, f_k \right\rangle$$

$$= \langle f, f_k \rangle - \sum_{j=1}^{n} \left\langle \langle f, f_j \rangle f_j, f_k \right\rangle$$

$$= \langle f, f_k \rangle - \sum_{j=1}^{n} \langle f, f_j \rangle \langle f_j, f_k \rangle$$

$$= \langle f, f_k \rangle - \sum_{j=1}^{n} \langle f, f_j \rangle \delta_{jk}$$

if $k \le n$
$$= \langle f, f_k \rangle - \langle f, f_k \rangle = 0.$$
if $k \le n$

This shows that $f - s_n$ is orthogonal to each f_k. Further,

$$\langle f - s_n, s_n \rangle = \left\langle f - s_n, \sum_{k=1}^{n} \langle f, f_k \rangle f_k \right\rangle$$

---

[3] Due to Friedrich Bessel (1784–1846; Westphalia, now Germany).

$$= \sum_{k=1}^{n} \left\langle f - s_n, \langle f, f_k \rangle f_k \right\rangle$$

$$= \sum_{k=1}^{n} \overline{\langle f, f_k \rangle} \langle f - s_n, f_k \rangle, \ \overset{?}{=} 0$$

which equals zero by the previous argument. This shows that $f - s_n$ is orthogonal to s_n. Then, by Exercise 4.2.1,

$$\|f - s_n\|^2 + \|s_n\|^2 = \|f\|^2.$$

This shows that

$$\|s_n\|^2 \le \|f\|^2.$$

Since

$$\|s_n\|^2 = \left\| \sum_{k=1}^{n} \langle f, f_k \rangle f_k \right\|^2,$$

which, by the same exercise and induction, is equal to

$$\sum_{k=1}^{n} \|\langle f, f_k \rangle f_k\|^2,$$

we have that

$$\|s_n\|^2 = \sum_{k=0}^{n} |\langle f, f_k \rangle|^2 \cdot \|f_k\|^2 = \sum_{k=1}^{n} |\langle f, f_k \rangle|^2.$$

Combining these last two sentences yields

$$\sum_{k=1}^{n} |\langle f, f_k \rangle|^2 \le \|f\|^2.$$

Since this holds for each n,

$$\sum_{k=1}^{\infty} |\langle f, f_k \rangle|^2 \le \|f\|^2,$$

as desired.   □

It is natural to want to determine conditions on $\{f_k\}_{k=1}^{\infty}$ under which equality in Bessel's inequality holds.

Theorem 4.3 (Parseval's Theorem[4]).   *As in the preceding theorem, suppose that* $\{f_k\}_{k=1}^{\infty}$ *is an orthonormal sequence in an inner product space* V. *Then* $\{f_k\}_{k=1}^{\infty}$ *is a complete orthonormal sequence if and only if for every* $f \in V$,

$$\sum_{k=1}^{\infty} |\langle f, f_k \rangle|^2 = \|f\|^2.$$

---

[4]Due to Marc-Antoine Parseval des Chênes (1755–1836; France).

PROOF.   This is left as Exercise 4.2.1(b).   □

We end this section with a sort of converse to Bessel's inequality. Theorem 4.2 implies, as a special case, that if $f \in L^2$, then the sum of squares of the Fourier coefficients of f, with respect to the trigonometric system $\{f_k\}_{k=1}^\infty$, is always finite. The combination of Theorems 4.2, 4.3, 4.4, and 4.6 sets up a "linear isometry" between $L^2([-\pi, \pi], m)$ and ℓ^2. Specifically, for $f \in L^2$, define $Tf = \{\langle f, f_k \rangle\}_{k=1}^\infty$, where $\{f_k\}_{k=1}^\infty$ denotes the trigonometric system. Then $Tf \in \ell^2$ (Theorem 4.2), T is linear, one-to-one (Theorem 4.6, together with Theorem 4.5 (c)), onto (Theorems 4.4 and 4.6), and $\|f\|_{L^2} = \|Tf\|_{\ell^2}$ (Theorem 4.3), for all $f \in L^2$. This result, that L^2 and ℓ^2 are isometrically isomorphic, is referred to as the Riesz–Fischer Theorem (Theorem 4.6 sometimes goes by the same name).

Theorem 4.4.   *Assume that*

(a) $\{d_k\}_{k=1}^\infty$ *is a sequence of real numbers such that* $\sum_{k=1}^\infty d_k^2$ *converges, and*
(b) *V is a Hilbert space with complete orthonormal sequence* $\{f_k\}_{k=1}^\infty$.

Then there is an element $f \in V$ whose Fourier coefficients with respect to $\{f_k\}_{k=1}^\infty$
are the numbers d_k, and

$$\|f\|^2 = \sum_{k=1}^\infty d_k^2.$$

PROOF.   Define

$$s_n = \sum_{k=1}^n d_k f_k.$$

For $m > n$,

$$\|s_n - s_m\|^2 = \sum_{j=n+1}^m \sum_{k=n+1}^m d_j d_k \langle f_j, f_k \rangle = \sum_{k=n+1}^m d_k^2.$$

Therefore, $\{s_n\}_{n=1}^\infty$ is Cauchy. Because V is a Hilbert space, there is an $f \in V$ such that

$$\lim_{n \to \infty} \|s_n - f\| = 0.$$

This is what we mean when we write

$$f = \sum_{k=1}^\infty d_k f_k,$$

and Theorem 4.1 now says that $d_k = \langle f, f_k \rangle$.
The remaining identity now follows from Parseval's theorem.   □

In order for the Riesz–Fischer theorem to be true, we would need to know that the trigonometric system is, in fact, a *complete* orthonormal sequence in $L^2([-\pi, \pi], m)$. This is the goal of the next section.

4.3   A Return to Classical Fourier Analysis

We now return to the classical setting, and the orthonormal family

$$\frac{1}{\sqrt{2\pi}}, \quad \frac{\cos(nx)}{\sqrt{\pi}}, \quad \frac{\sin(mx)}{\sqrt{\pi}}, \quad n, m = 1, 2, \ldots,$$

in $L^2([-\pi, \pi], m)$.

Our first theorem of this section is proved mainly for its use in the proof of Theorem 4.6 (that is why it appears here and not in the preceding section), but it is interesting in its own right. Parseval's theorem gives an alternative way to think about "completeness" of an orthonormal family; this theorem gives a few more ways. We state it only for orthonormal families in the specific Hilbert space $L^2([-\pi, \pi], m)$; the result can be generalized to arbitrary Hilbert spaces.

Theorem 4.5.   *For an orthonormal sequence $\{f_k\}_{k=1}^{\infty}$ in $L^2([-\pi, \pi], m)$, the following are equivalent:*

(a) $\{f_k\}_{k=1}^{\infty}$ *is a complete orthonormal sequence.*
(b) *For every $f \in L^2$ and $\epsilon > 0$ there is a finite linear combination*

$$g = \sum_{k=1}^{n} d_k f_k$$

such that $\| f - g \|_2 \leq \epsilon$.
(c) *If the Fourier coefficients with respect to $\{f_k\}_{k=1}^{\infty}$ of a function in L^2 are all 0, then the function is equal to 0 almost everywhere.*

PROOF.   It should be clear from the definition that (a) implies (b).

To prove that (b) implies (c), let f be a square integrable function such that $\langle f, f_k \rangle = 0$ for all k. Let $\epsilon > 0$ be given and choose g as in (b). Then

$$\| f \|_2^2 = \left| \| f \|_2^2 - \left\langle f, \sum_{k=1}^{n} d_k f_k \right\rangle \right| = |\langle f, f - g \rangle|$$

$$\leq \| f \|_2 \cdot \| f - g \|_2 \leq \epsilon \| f \|_2.$$

This implies that $\| f \|_2 \leq \epsilon$. Since ϵ was arbitrary, f must be 0 almost everywhere.

To prove that (c) implies (a), let $f \in L^2$ and put

$$s_n = \sum_{k=1}^{n} \langle f, f_k \rangle f_k.$$

As in the proof of Theorem 4.4, we see that $\{s_n\}_{n=1}^{\infty}$ is Cauchy in L^2. And Theorem 3.21 then tells us that there is a function $g \in L^2$ such that

$$\lim_{n \to \infty} \| s_n - g \|_2 = 0.$$

That is,

$$g = \sum_{k=1}^{\infty} \langle f, f_k \rangle f_k.$$

Theorem 4.1 then tells us that the Fourier coefficients of g are the same as the Fourier coefficients of f with respect to $\{f_k\}_{k=1}^{\infty}$, i.e., $\langle g, f_k \rangle = \langle f, f_k \rangle$. By (b), $f - g$ must equal 0 almost everywhere. In other words,

$$f = \sum_{k=1}^{\infty} \langle f, f_k \rangle f_k.$$

Since f was arbitrary, $\{f_k\}_{k=1}^{\infty}$ is a complete orthonormal sequence. □

Theorem 4.6. *The trigonometric system*

$$\frac{1}{\sqrt{2\pi}}, \quad \frac{\cos(nx)}{\sqrt{\pi}}, \quad \frac{\sin(mx)}{\sqrt{\pi}}, \quad n, m = 1, 2, \ldots,$$

forms a complete orthonormal sequence in $L^2([-\pi, \pi], m)$. That is, if f is such that $|f|^2$ is Lebesgue integrable, then its (classical) Fourier series converges to f. The convergence is convergence in the norm $\| \cdot \|_2$, i.e.,

$$\lim_{n \to \infty} \left[\int_{-\pi}^{\pi} \left[f(x) - \left(\frac{a_0}{2} + \sum_{k=1}^{n} a_k \cos(kx) + \sum_{k=1}^{n} b_k \sin(kx) \right) \right]^2 dx \right] = 0.$$

This type of convergence is often called "in mean" convergence.

PROOF.   In Exercise 4.1.1 you are asked to prove that the trigonometric system is orthonormal. We complete the proof of the theorem by verifying that condition (c) of Theorem 4.5 holds. First consider the case that f is continuous and real-valued, and that $\langle f, f_k \rangle = 0$ for each f_k. If $f \neq 0$, we then know that there exists an x_0 at which $|f|$ achieves a maximum, and we may assume that $f(x_0) > 0$. Let δ be small enough to ensure that $f(x) > \frac{f(x_0)}{2}$ for all x in the interval $(x_0 - \delta, x_0 + \delta)$. Consider the function

$$t(x) = 1 + \cos(x_0 - x) - \cos(\delta).$$

This function is a finite linear combination of functions in the trigonometric system; such functions are called "trigonometric polynomials." It is straightforward to verify

  (i) $1 < t(x)$, for all x in $(x_0 - \delta, x_0 + \delta)$, and
 (ii) $|t(x)| \leq 1$ for all x outside of $(x_0 - \delta, x_0 + \delta)$.
      Since f is orthogonal to every member of the trigonometric system, f is orthogonal to every trigonometric polynomial and, in particular, is orthogonal to t^n for every positive integer n. This will lead us to a contradiction. Notice that

$$0 = \langle f, t^n \rangle = \int_{-\pi}^{\pi} f(x) t^n(x) dx$$

$$= \int_{-\pi}^{x_0 - \delta} f(x) t^n(x) dx + \int_{x_0 - \delta}^{x_0 + \delta} f(x) t^n(x) dx + \int_{x_0 + \delta}^{\pi} f(x) t^n(x) dx.$$

By (ii) above, the first and third integrals are bounded in absolute value for each n by $2\pi f(x_0)$. The middle integral, however, is greater than or equal to $\int_a^b f(x) t^n(x) dx$, where $[a, b]$ is any closed interval in $(x_0 - \delta, x_0 + \delta)$. Since t is continuous on $[a, b]$, we know that t achieves a minimum value, m, there. By (i) above, $m > 1$. Then

$$\int_a^b f(x)t^n(x)dx \geq \frac{f(x_0)}{2} \cdot m^n \cdot (b - a),$$

which grows without bound as $n \to \infty$. This contradicts the assumption that $0 = \langle f, t^n \rangle$ for all n. Thus, any continuous real-valued function that is orthogonal to every trigonometric polynomial must be identically 0.

If f is continuous but not real-valued, our hypothesis implies that

$$\int_{-\pi}^{\pi} f(x)e^{-ikx}dx = 0, \quad k = 0, \pm 1, \pm 2, \ldots,$$

and thus also that

$$\int_{-\pi}^{\pi} \overline{f(x)}e^{-ikx}dx = 0, \quad k = 0, \pm 1, \pm 2, \ldots.$$

If we add and subtract these two equations, we see that the real and imaginary parts of f are orthogonal to each of the members of the trigonometric system. By the first part of the proof, the real and imaginary parts of f are identically 0; hence f is identically 0.

Finally, we no longer assume that f is continuous. Define the continuous function

$$F(x) = \int_{-\pi}^{x} f(t)dt.$$

For now let us assume that $f_k(x) = \frac{\cos(kx)}{\sqrt{\pi}}$. Our hypothesis implies

$$0 = \int_{-\pi}^{\pi} f(x)\cos(kx)dx.$$

Integration by parts yields

$$\int_{-\pi}^{\pi} F(x)\sin(kx)dx = \frac{1}{k}\int_{-\pi}^{\pi} f(x)\cos(kx)dx = 0.$$

Similarly, we can show that

$$\int_{-\pi}^{\pi} F(x)\cos(kx)dx = 0.$$

We now have shown that F, and hence $F - C$ for every constant C, is orthogonal to each of the nonconstant members of the trigonometric system. We now take care of the member $\frac{1}{\sqrt{2\pi}}$. Let

$$C_0 = \frac{1}{2\pi}\int_{-\pi}^{\pi} F(x)dx.$$

Then $F - C_0$ is easily seen to be orthogonal to *every* member of the trigonometric system. Since F is continuous, $F - C_0$ is also continuous, and the first part of the proof shows that $F - C_0$ is identically 0. From this it follows that $f = F'$ is 0 almost everywhere. □

Is this theorem "good"? Note that $\| \cdot \|_2$-convergence does not necessarily imply either uniform or pointwise convergence (Exercise 4.1.4). With uniform convergence, for example, we know that we cannot get the same result because the partial sums of the Fourier series of f are always continuous functions, and if the convergence of the series were uniform, then f would have to be continuous, too. Since $L^2([-\pi, \pi], m)$ contains discontinuous functions, we see that uniform convergence cannot always be achieved.

Theorem 4.6 has an important corollary, which we state as our next theorem. See Exercise 3.6.8 for an alternative proof of this same result.

Theorem 4.7. $C([-\pi, \pi])$ *is dense in* L^2.

PROOF. This is immediate, since the trigonometric polynomials are each continuous, and Theorem 4.6 shows that the smaller set is dense (see Theorem 4.5(b)). □

This theorem gives us an alternative way to define L^2. First, it is not hard to see that the interval $[-\pi, \pi]$ can be replaced by any other closed and bounded interval $[a, b]$. One can define $L^2([a, b], m)$ as the *completion* of $C([a, b])$ with respect to the norm $\| \cdot \|_2$. The advantage of this definition is that it gives a way of discussing the very important Hilbert space L^2 without ever mentioning general measure and integration theory. Specifically, we define L^2 to be the collection of functions f defined on the interval $[a, b]$ such that

$$\lim_{n \to \infty} \| f_n - f \|_2 = 0$$

for some sequence $\{f_n\}_{n=1}^\infty \in C([a, b])$. Actually, L^2 must be considered to be the equivalence classes of such functions, where two functions are equivalent if and only if they are equal almost everywhere (see the discussion preceding Theorem 3.17). Therefore, it is not entirely true that this definition avoids discussing measure. However, we can give this definition with only an understanding of "measure zero," and not general measure. (And measure zero can be defined in a straightforward manner and is much simpler to understand than general measure.)

Exercises for Chapter 4

Section 4.1

4.1.1 Show that the trigonometric system

$$\frac{1}{\sqrt{2\pi}}, \quad \frac{\cos(nx)}{\sqrt{\pi}}, \quad \frac{\sin(mx)}{\sqrt{\pi}}, \quad n, m = 1, 2, \ldots,$$

is an orthonormal sequence in $L^2([-\pi, \pi], m)$.

4.1.2 In this exercise you will actually compute a classical (i.e., with respect to the orthonormal sequence of Exercise 1) Fourier series, and investigate its convergence properties. The function given is a basic one; in the next exercise you are asked to do the same procedure with another very basic function. You are being asked to do these by hand, and you can no doubt appreciate that the computations get quite laborious once we depart from even the most basic functions. There are tricks for doing these computations; the interested reader can learn more about such techniques in a text devoted to classical Fourier series.

(a) Let

$$f(x) = \begin{cases} 1 & \text{if } -\pi \leq x < 0, \\ 0 & \text{if } 0 \leq x < \pi. \end{cases}$$

Show that its Fourier series is

$$\frac{1}{2} + \frac{1}{\pi} \sum_{n=1}^{\infty} \frac{((-1)^n - 1)}{n} \sin(nx).$$

(b) Explain why this series converges in mean to f.

(c) What can you say about the *pointwise* and *uniform* convergence of this series?

(d) Why are the coefficients of the cosine terms all zero?

4.1.3 In this exercise you will compute another Fourier series and investigate its convergence properties.

(a) Let $f(x) = x^2$. Show that its classical Fourier series is

$$\frac{\pi^2}{3} + 4 \sum_{n=1}^{\infty} \frac{(-1)^n}{n^2} \cos(nx).$$

(b) Explain why this series converges in mean to f.

(c) What can you say about the *pointwise* and *uniform* convergence of this series?

(d) Why are the coefficients of the sine terms all zero?

4.1.4 For a sequence $\{f_n\}_{n=1}^{\infty}$ in $L^2([-\pi, \pi], m)$, we have seen three ways for $\{f_n\}_{n=1}^{\infty}$ to converge:

(i) "pointwise,"
(ii) "uniformly,"
(iii) "in mean."

The point of this exercise is to understand the relation between these three types of convergence. For the counterexamples asked for below, use whatever finite interval $[a, b]$ you find convenient. Please make an effort to supply "easy" examples.

(a) Prove that uniform convergence implies pointwise convergence.

(b) Give an example to show that pointwise convergence does not imply uniform convergence.

(c) Prove that uniform convergence implies convergence in mean.

(d) Give an example to show that pointwise convergence does not imply convergence in mean.

(e) Give an example to show that convergence in mean does not imply pointwise convergence. (Note that the same example shows that convergence in mean does not imply uniform convergence.)

4.1.5 Apply the Gram–Schmidt process to the functions $1, x, x^2, x^3, \ldots$ to obtain formulas for the first three Legendre polynomials. Then verify that they are indeed given by the formula

$$\sqrt{\frac{2n+1}{2}}\,\frac{1}{2^n n!}\,\frac{d^n}{dx^n}(x^2-1)^n, \qquad n = 1, 2, 3.$$

4.1.6 Prove that the Haar family is an orthonormal family in the Hilbert space $L^2([0, 1])$.

4.1.7 **(a)** Show that the sequence

$$\frac{e^{inx}}{\sqrt{2\pi}}, \qquad n = 0, \pm 1 \pm 2, \ldots,$$

is a complete orthonormal sequence in $L^2([-\pi, \pi])$.

(b) Show that the sequence

$$\sqrt{\frac{2}{\pi}}\cos(nx), \qquad n = 1, 2, 3, \ldots,$$

is a complete orthonormal sequence in $L^2([0, \pi])$. (Observe that $\sqrt{\frac{2}{\pi}}\cos(nx)$ can be replaced by $\sqrt{\frac{2}{\pi}}\sin(nx)$.)

Section 4.2

4.2.1 **(a)** Prove that in any inner product space $(V, \langle \cdot, \cdot \rangle)$, f and g orthogonal implies

$$\|f\|^2 + \|g\|^2 = \|f + g\|^2.$$

Here, as usual, $\|\cdot\| = \sqrt{\langle \cdot, \cdot \rangle}$.

(b) Prove Parseval's theorem.

4.2.2 Assume that $\{f_n\}_{n=1}^\infty$ is a sequence in L^2 and that $f_n \to f$ in mean. Prove that $\{\|f_n\|_2\}_{n=1}^\infty$ is a bounded sequence of real numbers.

4.2.3 Assume that $f_1, f_2, \ldots, f_n$ is an orthonormal family in an inner product space. Prove that $f_1, f_2, \ldots, f_n$ are linearly independent.

4.2.4 For f and g in an inner product space, $g \neq 0$, the *projection of f on g* is the vector

$$\frac{\langle f, g \rangle}{\|g\|^2}\,g.$$

Show that the two vectors

$$\frac{\langle f, g \rangle}{\|g\|^2} g \quad \text{and} \quad f - \frac{\langle f, g \rangle}{\|g\|^2} g$$

are orthogonal.

4.2.5 **(a)** Show that the classical Fourier series of $f(x) = x$ is

$$2 \sum_{n=1}^{\infty} \frac{(-1)^{n+1}}{n} \sin(nx).$$

(b) Use your work in (a), together with Parseval's identity, to obtain Euler's remarkable identity

$$\sum_{n=1}^{\infty} \frac{1}{n^2} = \frac{\pi^2}{6}.$$

(Note: The same procedure can be applied to the Fourier series of x^2 to obtain

$$\sum_{n=1}^{\infty} \frac{1}{n^4} = \frac{\pi^4}{90},$$

and so on!)

5

An Introduction to Abstract Linear Operator Theory

In this chapter you will read about the beginning material of operator theory. The chapter is written with the aim of getting to spectral theory as quickly as possible. Matrices are examples of linear operators. They transform one linear space into another and do so linearly. "Spectral values" are the infinite-dimensional analogues of eigenvalues in the finite-dimensional situation. Spectral values can be used to decompose operators, in much the same way that eigenvalues can be used to decompose matrices. You will see an example of this sort of decomposition in the last section of this chapter, where we prove the spectral theorem for compact Hermitian operators. One of the most important open problems in operator theory at the start of the twenty-first century is the "invariant subspace problem." In the penultimate section of this chapter we give a description of this problem and discuss some partial solutions to it. We also let the invariant subspace problem serve as our motivation for learning a bit about operators on Hilbert space. The material found at the end of Section 3 (from Theorem 5.7 onwards) through the last section (Section 5) of the chapter is not usually covered in an undergraduate course. This material is sophisticated, and will probably seem more difficult than other topics we cover.

Further basic linear operator theory can be found in Section 6.3.

5.1 Basic Definitions and Examples

We start by considering two real (or complex) linear spaces, X and Y. A mapping T from X to Y is called a *linear operator* if the domain of T, D_T, is a linear subspace of X, and if

$$T(\alpha x + \beta y) = \alpha Tx + \beta Ty$$

for all $x, y \in D_T$ and all real (or complex) scalars α and β. Notice that any linear map satisfies $T(0) = 0$. In this context it is common to write Tx in place of $T(x)$, and unless otherwise stated, D_T is taken to be all of X. The first thing we want to do is establish a reasonable list of examples of linear operators.

EXAMPLE 1.   As should be familiar from linear algebra, any real $m \times n$ matrix (a_{ij}) defines a linear operator from $\mathbb{R}^n$ to $\mathbb{R}^m$ via

$$\begin{pmatrix} a_{11} & a_{12} & \cdots & a_{1n} \\ a_{21} & a_{22} & \cdots & a_{2n} \\ \vdots & \vdots & \ddots & \vdots \\ a_{m1} & a_{m2} & \cdots & a_{mn} \end{pmatrix} \begin{pmatrix} x_1 \\ \vdots \\ x_n \end{pmatrix} = \left(\sum_{j=1}^{n} a_{1j}x_j, \sum_{j=1}^{n} a_{2j}x_j, \ldots, \sum_{j=1}^{n} a_{mj}x_j \right).$$

EXAMPLE 2.   An "infinite matrix" (a_{ij}), $i, j = 1, 2, \ldots$, can be used to represent an operator on a sequence space. For example, the infinite matrix

$$\begin{pmatrix} 0 & 1 & 0 & 0\ldots \\ 0 & 0 & 1 & 0\ldots \\ 0 & 0 & 0 & 1\ldots \\ \vdots & \vdots & \vdots & \ddots \end{pmatrix}$$

represents the linear operator T acting on $\ell^2 = \ell^2(\mathbb{N})$ (or ℓ^∞, etc.) given by

$$T(x_1, x_2, \ldots) = (x_2, x_3, \ldots).$$

This is an example of what is called a *shift operator*. More specifically, it is likely to be referred to as the "backward unilateral shift," or "unilateral shift," or "left shift." We will use the last of these names. Another important shift is the "right shift," defined by

$$S(x_1, x_2, \ldots) = (0, x_1, x_2, \ldots).$$

What is the (infinite) matrix representing this shift? There are also "weighted shifts." For example, the sequence of 1's in the matrix of T can be replaced by a suitable sequence $\{a_i\}_{i=1}^{\infty}$ of scalars, and the weighted shift thus constructed sends $(x_1, x_2, \ldots)$ to $(a_1 x_2, a_2 x_3, \ldots)$. The class of shift operators plays an important role in the theory of operators on Hilbert spaces.

EXAMPLE 3.   For our first example of a linear operator on a function space, we observe that the map

$$Tf = \int_0^1 f(t)dt$$

defines a linear operator $T : C([a, b]) \to \mathbb{R}$.

Integral and differential equations are rich areas of application for operator theory, and they provided impetus for the early development of functional analysis. Many important classes of linear operators involve integrals.

EXAMPLE 4.   The map

$$(Tf)(s) = \int_a^b k(s,t)f(t)dt$$

defines a linear operator $T : C([a,b]) \to C([a,b])$, where $k(s,t)$ is defined for $a \le s \le b, a \le t \le b$ and, for each function $f \in C([a,b])$, the function

$$s \to \int_a^b k(s,t)f(t)dt$$

is continuous on $[a,b]$. This is called a *Fredholm operator of the first kind*, named in honor of Erik Ivar Fredholm (1866–1927; Sweden). These operators are an infinite-dimensional version of the first example: Imagine that the variables s and t take on integer values only, and that the function $k(s,t)$ is the matrix (k_{st}).

EXAMPLE 5.   Let k be as in Example 4. A slight variation on the preceding example leads us to the linear operator

$$(Tf)(s) = f(s) - \int_a^b k(s,t)f(t)dt.$$

This is called a *Fredholm operator of the second kind*.

If $k(s,t) = 0$ for all $t > s$ (that is, k is "lower triangular"), the operators of Examples 4 and 5 are called *Volterra-type operators* (of the first and second kind, respectively). Vito Volterra was a powerful mathematician; we have already encountered some of his many contributions. It is his work on integral equations that served as impetus for some of the early development of functional analysis. Already by the age of thirteen, Volterra was working on the three body problem. Throughout his life he was a strong promoter of international collaboration among scientists, and he traveled extensively to support this cause. He refused to take the oath of allegiance to the Italy's Fascist government and so, in 1931, was forced to leave his position at the University of Rome. He spent most of the rest of his life in Paris.

The formulas of Examples 4 and 5 can be used to define integral operators on different function spaces. For example, if $k(s,t) \in L^2([a,b] \times [a,b])$, then the formulas define linear operators from $L^2([a,b])$ to itself. You will get to work with an operator of this type in Exercise 5.3.12.

EXAMPLE 6.   We end with one more example, of a fundamental kind. If H is a Hilbert space with inner product $\langle \cdot, \cdot \rangle$ and x_0 is a designated element of H, then the map $Tx = \langle x, x_0 \rangle$ defines a linear operator $T : H \to \mathbb{C}$. These are examples of what are called "linear functionals" and are much more important than will be made clear in this book. See Section 6.3 for more on linear functionals.

When X and Y are normed spaces, as they are in all our examples above, a linear operator $T : X \to Y$ may or may not be continuous. One very nice consequence of being linear is that continuity need be checked only at a single point. This is what our first theorem of the chapter says.

Theorem 5.1. *Let X and Y be normed linear spaces. A linear operator $T:X \to Y$ is continuous at every point if it is continuous at a single point.*

PROOF.   Suppose that T is continuous at the point x_0, and let x be any point in X and $\{x_n\}_{n=1}^{\infty}$ a sequence in X converging to x. Then the sequence $\{x_n - x + x_0\}_{n=1}^{\infty}$ converges to x_0, and therefore, since T is continuous at x_0, $\{T(x_n - x + x_0)\}_{n=1}^{\infty}$ converges to Tx_0. Since T is linear,

$$T(x_n - x + x_0) = Tx_n - Tx + Tx_0,$$

and hence $\{Tx_n\}_{n=1}^{\infty}$ converges to Tx. Since x was chosen arbitrarily, T is continuous on all of X.   □

By this theorem, in order to see that a given linear operator is continuous, it suffices to check continuity at 0.

5.2   Boundedness and Operator Norms

Let X and Y be normed linear spaces. A linear operator $T : X \to Y$ is said to be *bounded* if there exists an $M > 0$ such that

$$\|Tx\|_Y \le M\|x\|_X$$

for all $x \in X$.

We shall now discuss the boundedness of a few of the operators introduced in the previous section.

EXAMPLE 1.   The left shift $T : \ell^2 \to \ell^2$ defined by

$$T(x_1, x_2, \ldots) = (x_2, x_3, \ldots)$$

is bounded because

$$\|T(x_1, x_2, \ldots)\|_{\ell^2} = \sum_{k=2}^{\infty} |x_k|^2 \le \sum_{k=1}^{\infty} |x_k|^2 = \|(x_1, x_2, \ldots)\|_{\ell^2}.$$

We may choose $M = 1$ in this case.

EXAMPLE 2.   Let $k(s, t)$ be a continuous function of two variables defined for all $s, t \in [a, b]$. The integral operator $T : C([a, b]) \to C([a, b])$ defined by

$$(Tf)(s) = \int_a^b k(s, t)f(t)dt$$

is thus a Fredholm operator of the first kind. It is bounded because

$$\|Tf\|_\infty = \max_{a\le s\le b}\left\{\left|\int_a^b k(s,t)f(t)dt\right|\right\}$$

$$\le \max_{a\le s\le b}\left\{\int_a^b |k(s,t)|\cdot|f(t)|dt\right\}$$

$$\le \max_{a\le s\le b}\left\{\int_a^b \left(\max_{a\le t\le b}|k(s,t)|\right)\left(\max_{a\le t\le b}|f(t)|\right)dt\right\}$$

$$= \max_{a\le s\le b}\left\{\left(\max_{a\le t\le b}|k(s,t)|\right)\cdot\|f\|_\infty(b-a)\right\}$$

$$= \left(\max_{a\le s\le b}\max_{a\le t\le b}|k(s,t)|\right)\cdot(b-a)\cdot\|f\|_\infty.$$

We may choose $M = (\max_{a\le s\le b}\max_{a\le t\le b}|k(s,t)|)\cdot(b-a)$ in this case.

In the first example, any number larger than 1 can also be used for M, but no number smaller may be used (you can see this by considering, for example, $(x_1, x_2, \ldots) = (0, 1, 0, \ldots)$). In other words, 1 is the smallest number M such that

$$\|Tx\|_{\ell^2} \le M\|x\|_{\ell^2}$$

for all $x \in X$. Since this is the case, we say that the left shift operator $\ell^2 \to \ell^2$ has "operator norm" 1, and write

$$\|T\|_{\mathcal{B}(\ell^2)} = 1.$$

(The notation $\mathcal{B}(\ell^2)$ will be explained very shortly.) In general, the *norm* of a bounded operator $T : X \to Y$ is defined to be

$$\inf\{M \mid \|Tx\|_Y \le M\|x\|_X\}.$$

The collection of all bounded linear operators from X to Y will be denoted by $\mathcal{B}(X, Y)$. If $X = Y$, we follow the standard practice and write $\mathcal{B}(X)$ for $\mathcal{B}(X, Y)$. The norm of T considered as an operator from X to Y is denoted by $\|T\|_{\mathcal{B}(X,Y)}$, or simply by $\|T\|$ if the spaces X and Y are clear from context. Beware: The operator norm depends on the norms on the spaces X and Y. For example, the integral operator given by the formula

$$(Tf)(s) = \int_0^1 k(s,t)f(t)dt,$$

with suitable conditions on the function $k(s,t)$, can be considered either as an element of $\mathcal{B}(C([0,1]))$ or as an element of $\mathcal{B}(L^2([0,1]))$. In this case, the operator norms

$$\|T\|_{\mathcal{B}(C([0,1]))} \qquad \text{and} \qquad \|T\|_{\mathcal{B}(L^2([0,1]))}$$

need not be equal. See also Exercise 5.2.6.

If $x = 0$ in the inequality

$$\|Tx\|_Y \le M\|x\|_X,$$

then any number M works, so we may assume that $x \neq 0$. We thus see that the norm of T may also be defined to be

$$\sup\left\{\frac{\|Tx\|_Y}{\|x\|_X} \; \middle| \; x \neq 0\right\}.$$

Further,

$$\left\{\frac{\|Tx\|_Y}{\|x\|_X} \; \middle| \; x \neq 0\right\} = \left\{\left\|T\left(\frac{x}{\|x\|_X}\right)\right\|_Y \; \middle| \; x \neq 0\right\}$$
$$\subseteq \{\|Tx\|_Y \mid \|x\|_X = 1\}$$
$$\subseteq \{\|Tx\|_Y \mid \|x\|_X \leq 1\}.$$

Put

$$M_1 = \sup\left\{\frac{\|Tx\|_Y}{\|x\|_X} \; \middle| \; x \neq 0\right\},$$
$$M_2 = \sup\{\|Tx\|_Y \mid \|x\|_X = 1\},$$
$$M_3 = \sup\{\|Tx\|_Y \mid \|x\|_X \leq 1\}.$$

The above set inclusions show that

$$M_1 \leq M_2 \leq M_3.$$

However, if $x \neq 0$, then we have that

$$\frac{\|Tx\|_Y}{\|x\|_X} \leq M_1$$

and hence

$$\|Tx\|_Y \leq M_1\|x\|_X.$$

If, further, $\|x\|_X \leq 1$, then

$$\|Tx\|_Y \leq M_1\|x\|_X \leq M_1.$$

Taking the supremum now over all $\|x\|_X \leq 1$ yields

$$M_3 \leq M_1.$$

Thus, in fact,

$$M_1 = M_2 = M_3.$$

This gives a few (slightly) different ways of thinking about the norm of an operator.

How does one compute an operator norm? To try to calculate $\|T\|$ for a specific operator T, first establish an upper bound for $\|T\|$ and try to make it is small as possible. This is done by thinking of the least value of M that makes

$$\|Tx\| \leq M\|x\|$$

valid for every x in the domain space. Then try to show that this value for M cannot be improved upon by picking an element x (or sequence $\{x_n\}_{n=1}^{\infty}$) for which this bound is attained (or is the supremum).

Theorem 5.2. *Let X and Y be normed linear spaces. A linear operator $T : X \to Y$ is continuous on X if and only if it is bounded on X.*

PROOF.   First assume that T is continuous on X. Then it is continuous at 0 and thus there exists a $\delta > 0$ such that $\|x\|_X \leq \delta$ implies

$$\|Tx\|_Y = \|Tx - T0\|_Y \leq 1.$$

Let $x \in X$ be arbitrary and set $x_0 = \frac{\delta x}{\|x\|_X}$. Then

$$1 \geq \|Tx_0\|_Y = \frac{\delta}{\|x\|_X} \|Tx\|_Y,$$

showing that

$$\|Tx\|_Y \leq \frac{1}{\delta} \|x\|_X$$

for all $x \in X$. In other words, T is bounded, and the norm of T is at most $\frac{1}{\delta}$.

For the other direction, it suffices to show that T is continuous at 0. Let $\{x_n\}_{n=1}^{\infty}$ be a sequence converging to 0. Then, since T is bounded,

$$\|Tx_n\|_Y \leq \|T\| \cdot \|x_n\|_X$$

for all n, and hence $\|Tx_n\|_Y \to 0$ as $n \to \infty$ (that is, T is continuous at 0). $\square$

Theorem 5.3. *Let X and Y be normed linear spaces. The collection $\mathcal{B}(X, Y)$ of all bounded linear operators $T : X \to Y$, endowed with the operator norm as discussed above, forms a normed linear space.*

PROOF.   Left as Exercise 5.2.1. $\square$

Theorem 5.4. *Let X and Y be normed linear spaces. $\mathcal{B}(X, Y)$ is a Banach space whenever Y is a Banach space.*

PROOF.   By Theorem 5.3 all that remains to be shown is that $\mathcal{B}(X, Y)$ is complete whenever Y is complete. Let $\{T_n\}_{n=1}^{\infty}$ be a Cauchy sequence in $\mathcal{B}(X, Y)$. We aim to show that $\{T_n\}_{n=1}^{\infty}$ converges.

Let $\epsilon > 0$. Then there exists an N such that

$$\|T_n - T_m\|_{\mathcal{B}(X,Y)} < \epsilon$$

whenever $n, m \geq N$. Also, since any Cauchy sequence is bounded, there exists $M > 0$ such that $\|T_n\|_{\mathcal{B}(X,Y)} \leq M$ for all n.

Now, $\{T_n\}_{n=1}^{\infty}$ Cauchy implies that the sequence $\{T_n x\}_{n=1}^{\infty}$ is Cauchy for each $x \in X$. Since Y is assumed complete, $\{T_n x\}_{n=1}^{\infty}$ converges for each x. Let Tx denote the limit of the sequence $\{T_n x\}_{n=1}^{\infty}$. The operator T thus defined is linear, and

$$\|Tx\|_Y \leq M \|x\|_X$$

for all $x \in X$ (that is, T is a bounded operator). Therefore, $T \in \mathcal{B}(X, Y)$. T is our candidate for the limit of the sequence $\{T_n\}_{n=1}^{\infty}$ in $\mathcal{B}(X, Y)$.

For $n > m \geq N$,

$$\|T_n - T_m\|_{B(X,Y)} < \epsilon,$$

and so

$$\|T_n x - T_m x\|_Y < \epsilon \|x\|_X \tag{5.1}$$

for all $x \in X$. If we hold m fixed, then

$$\|T_n x - T_m x\|_Y \to \|T x - T_m x\|_Y$$

as $n \to \infty$. Therefore, letting $n \to \infty$ in (5.1) yields

$$\|T x - T_m x\|_Y \leq \epsilon \|x\|_X$$

for all $x \in X$. In other words,

$$\|T - T_m\|_{B(X,Y)} \leq \epsilon$$

for all $m \geq N$, and the proof is complete. □

5.3 Banach Algebras and Spectra; Compact Operators

An *algebra* is a linear space $\mathcal{A}$ together with a definition of multiplication satisfying four conditions:

(i) $a(bc) = (ab)c$,
(ii) $a(b + c) = ab + ac$,
(iii) $(a + b)c = ac + bc$,
(iv) $\lambda(ab) = (\lambda a)b = a(\lambda b)$

for $a, b, c \in \mathcal{A}$ and scalar λ.

The algebra is called real or complex according to whether the scalar field is the real or complex numbers. *For the remainder of the chapter, and unless specifically mentioned, all scalars will be assumed complex.* If there is an element $e \in \mathcal{A}$ such that $ea = ae = a$ for all $a \in \mathcal{A}$, we have a *unital algebra*.

An algebra $\mathcal{A}$ that is also a normed linear space whose norm is *submultiplicative*, that is, it satisfies

$$\|ab\| \leq \|a\| \cdot \|b\|$$

for all $a, b \in \mathcal{A}$, is called a *normed algebra*. If the norm on a normed algebra is complete, then $\mathcal{A}$ is a *Banach algebra*.

Assume X is a normed linear space and consider $B(X) = B(X, X)$. This, by Theorems 5.3 and 5.4, is a normed linear space and is a Banach space whenever X is. Notice, too, that $S \circ T \in B(X)$ whenever $S, T \in B(X)$. This property, together with the fact that $B(X)$ is a linear space, makes $B(X)$ into an algebra. In addition, note that $\|S \circ T\| \leq \|S\| \cdot \|T\|$ for all $S, T \in B(X)$ (you are asked to do this in Exercise 5.3.1). Thus, $B(X)$ is a Banach algebra whenever X itself is a Banach space. If we define $I \in B(X)$ by $Ix = x$ for all $x \in X$, then I serves as a multiplicative identity for $\mathcal{B}(X)$. Hence $\mathcal{B}(X)$ is a unital Banach algebra. We usually write ST in place of $S \circ T$.

An element S of $\mathcal{B}(X)$ is called *invertible* if there exists $T \in \mathcal{B}(X)$ such that $ST = I = TS$. It is important to realize that S may have a linear space inverse, T, on X without being invertible in $\mathcal{B}(X)$. In this case, T will fail to be a bounded (continuous) operator. If X is a Banach space, then T must always be bounded (Exercise 6.3.1). Along these same lines, note that S may have a "left inverse," yet not be invertible. Consider, for example, the right shift on $\ell^2 \to \ell^2$ given by

$$S(x_1, x_2, \ldots) = (0, x_1, x_2, \ldots).$$

The left shift

$$T(x_1, x_2, \ldots) = (x_2, x_3 \ldots)$$

satisfies $TS = I$ and hence serves as a "left inverse" for S. There is, however, no "right inverse" for S. It is interesting to observe that this behavior cannot happen for finite matrices. That is, if an $n \times n$ matrix A has a right inverse B ($AB = I$), then B must also be a left inverse for A. See Exercise 5.1.3.

Sometimes, the inverse of a given operator is obvious. Sometimes, it is equally obvious that a given operator is not invertible. For example, it may be clear that the operator is not one-to-one, as is the case with the left shift. Likewise, it may be clear that an operator is not onto, as is the case with the right shift. it would be nice to have some "tests" for invertibility. Preferably, a test would be easy to use. The next theorem gives one such test.

Theorem 5.5. *Let X be a Banach space. Suppose that $T \in \mathcal{B}(X)$ is such that $\|T\| < 1$. Then the operator $I - T$ is invertible in $\mathcal{B}(X)$, and its inverse is given by*

$$(I - T)^{-1} = \sum_{k=0}^{\infty} T^k.$$

PROOF.   Exercise 5.3.5 tells us that $\|T^k\| \leq \|T\|^k$ for each positive integer k. Therefore,

$$\sum_{k=0}^{\infty} \|T^k\| \leq \sum_{k=0}^{\infty} \|T\|^k,$$

and Lemma 3.20 and Theorem 5.4 together tell us that the series $\sum_{k=0}^{\infty} T^k$ is an element of $\mathcal{B}(X)$. It remains to be shown that $S = \sum_{k=0}^{\infty} T^k$ is the inverse for $I - T$. Since $I - T$ is continuous, we have

$$(I - T)Sx = (I - T)\left(\lim_{n \to \infty} \sum_{k=0}^{n} T^k \right)x$$

$$= \left(\lim_{n \to \infty} \sum_{k=0}^{n} (I - T)T^k \right)x$$

$$= \lim_{n \to \infty} \left(x - T^{n+1}x \right)$$

$$= x - \lim_{n \to \infty} T^{n+1} x$$

for each $x \in X$. The result of Exercise 5.3.5 implies that $\lim_{n \to \infty} T^{n+1} x = 0$ and therefore $(I - T)Sx = x$ for each $x \in X$. In a similar fashion, we can show that $S(I - T)x = x$ for each $x \in X$, and so $(I - T)^{-1} = \sum_{k=0}^{\infty} T^k$, as desired.   □

As a corollary to this theorem, we get the following result.

Theorem 5.6. *Let X be a Banach space. Suppose that $S, T \in \mathcal{B}(X)$, T is invertible, and $\|T - S\| < \|T^{-1}\|^{-1}$. Then S is invertible in $\mathcal{B}(X)$.*

PROOF.   Observe that if A and B are two invertible elements of $\mathcal{B}(X)$, then their product is also invertible and $(AB)^{-1} = B^{-1}A^{-1}$. If S and T satisfy the hypotheses of the theorem, then

$$\|(T - S)T^{-1}\| \leq \|T - S\| \cdot \|T^{-1}\| < 1.$$

The preceding theorem then shows that $I - (T - S)T^{-1} = ST^{-1}$ is invertible. By the observation made at the beginning of this proof, $S = (ST^{-1})T$ is invertible.   □

Let $\mathcal{G}$ denote the set of all invertible elements in $\mathcal{B}(X)$. An important corollary of Theorem 5.6 is the fact that $\mathcal{G}$ is an open subset of $\mathcal{B}(X)$. Theorems 5.5 and 5.6, as well as the fact that the set of invertible elements is open (and also Theorems 5.7 and 5.8), remain true when $\mathcal{B}(X)$ is replaced by an arbitrary unital Banach algebra.

Recall, from linear algebra, that an eigenvalue of an $n \times n$ matrix A is a scalar λ such that $\lambda I - A$ is *not* invertible. Here, I denotes the $n \times n$ identity matrix.

The *spectrum* of an element a in a unital Banach algebra $\mathcal{A}$ is defined to be the set of all complex numbers λ such that $\lambda e - a$ is not invertible in $\mathcal{A}$. The spectrum of a in $\mathcal{A}$ is denoted by $\sigma_{\mathcal{A}}(a)$, or by $\sigma(a)$ if there is no risk of confusion. "Spectral theory" is the study of this set. Not surprisingly, the name spectrum has physical interpretations. For example, in quantum mechanics, any "observable" has a (Hermitian) operator (on a Hilbert space) associated to it, and the observable can assume only values that appear in the spectrum of this operator. In Section 6.7 we will discuss the role of operator theory in quantum mechanics. Hilbert himself coined the phrase "spectral theory" in the context of his study of Fredholm's integral operators. However, he did not know that his spectra could have applications to physics. Indeed, Hilbert claimed, "I developed my theory of infinitely many variables from purely mathematical interests, and even called it 'spectral analysis' without any presentiment that it would later find an application to the actual spectrum of physics" [104].

Of particular interest to us is the spectrum of an operator T in the Banach algebra $\mathcal{B}(X)$.

For our first examples, we turn to the comforting setting of finite dimensions. We consider the matrices

$$A = \begin{pmatrix} 2 & 0 & 0 \\ 0 & 8 + 2i & 0 \\ 0 & 0 & 0 \end{pmatrix},$$

$$B = \begin{pmatrix} 4i & 1 & -1-i \\ 0 & 3 & 7 \\ 0 & 0 & 3+i \end{pmatrix},$$

$$C = \begin{pmatrix} 5 & 0 & 0 \\ 1 & 1 & 2 \\ 1 & -1 & -1 \end{pmatrix}.$$

as elements of $\mathcal{B}(\mathbb{C}^3)$. Then, as you should check,

$$\sigma(A) = \{0, 2, 8+2i\}, \qquad \sigma(B) = \{3, 4i, 3+i\}, \qquad \text{and} \qquad \sigma(C) = \{5, i, -i\}.$$

Given a finite subset of $\mathbb{C}$, how can you construct an operator with that specified set as its spectrum? What if the given set is countably infinite? Can these tasks always be achieved? If the set is finite, with n elements, we simply construct the $n \times n$ matrix with the elements of the set down the diagonal. This matrix will be an operator in $\mathcal{B}(\mathbb{C}^n)$ with the desired spectrum.

If the given set $\{\lambda_n\}_{n=1}^{\infty}$ is countably infinite, consider, by analogy, the diagonal matrix

$$D = \text{diag}(\lambda_1, \lambda_2, \ldots) = \begin{pmatrix} \lambda_1 & 0 & 0 & \cdots \\ 0 & \lambda_2 & 0 & \cdots \\ 0 & 0 & \lambda_3 & \cdots \\ \vdots & \vdots & \vdots & \ddots \end{pmatrix}$$

as an operator on $\ell^2 = \ell^2(\mathbb{N})$. We will now explore the possibility that the given set $\{\lambda_n\}_{n=1}^{\infty}$ is actually the spectrum of D. First, we must describe the sequences $\{\lambda_n\}_{n=1}^{\infty}$ for which the associated operator $D = \text{diag}(\lambda_1, \lambda_2, \ldots)$ is a bounded operator on ℓ^2. If the sequence is bounded, then, as one can readily check, $D \in \mathcal{B}(\ell^2)$. Is this also a necessary condition? Consider an unbounded sequence $\{\lambda_n\}_{n=1}^{\infty}$. We may assume that $|\lambda_n| \geq n$. Consider the element of ℓ^2 defined by $x_n = \frac{1}{\lambda_n}$; the image of this element is not even in ℓ^2. Therefore, $D = \text{diag}(\lambda_1, \lambda_2, \ldots)$ is in $\mathcal{B}(\ell^2)$ if and only if $\{\lambda_n\}_{n=1}^{\infty}$ is a bounded sequence. In this case we can at least consider the set $\sigma(D)$. Is $\sigma(D)$ equal to $\{\lambda_n\}_{n=1}^{\infty}$? For each n we see that the operator $\lambda_n I - D$ has nontrivial kernel and hence is not one-to-one and hence is not invertible. This establishes the containment $\{\lambda_n\}_{n=1}^{\infty} \subseteq \sigma(D)$. On the other hand, suppose that $\lambda \neq \lambda_n$ for all n. Then the matrix

$$\text{diag}((\lambda - \lambda_1)^{-1}, (\lambda - \lambda_2)^{-1}, \ldots)$$

serves as an inverse for $\lambda I - D$ as long as it defines a bounded operator, and this happens if and only if the sequence $\{(\lambda - \lambda_n)^{-1}\}_{n=1}^{\infty}$ is bounded. Since $\{(\lambda - \lambda_n)^{-1}\}_{n=1}^{\infty}$ is bounded if and only if $\{\lambda_n\}_{n=1}^{\infty}$ does not converge to λ, we see that $\sigma(D)$ is *not*, in fact, the sequence $\{\lambda_n\}_{n=1}^{\infty}$, but that it also contains its limit points. Indeed, it is impossible to construct a bounded linear operator with a spectrum that is not closed. This is a part of the next theorem (Theorem 5.7).

The next two theorems are fundamental in the theory of bounded operators. As mentioned at the beginning of the chapter, the material from this point on in this chapter becomes more sophisticated. Indeed, proofs of the next two theorems will not be given in full detail. To do so would take us too far afield into the study of functions defined on the complex plane. Nonetheless, we think it worthwhile to introduce this material now. We will give an indication of how the proofs of the next two theorems go, prove as much as we can, and point the reader in the direction he or she would need to go to complete the proofs. Before we discuss these proofs we do need to make a very brief digression into the theory of functions of a single complex variable. Let U be an open subset of the complex plane. A function $f : U \to \mathbb{C}$ is said to be *analytic* at $z_0 \in U$ if f can be represented by a power series centered at z_0, that is, if there is a positive number $r > 0$ and scalars $a_1, a_2, \ldots \in \mathbb{C}$ such that $|z - z_0| < r$ implies $z \in U$ and

$$f(z) = \sum_{n=0}^{\infty} a_n (z - z_0)^n$$

for all $|z - z_0| < r$. This is one of a few equivalent definitions of analyticity. In a first course on complex functions one studies analytic functions in great detail, and one encounters a remarkable result about their behavior due to Joseph Liouville (1809–1882; France). Liouville's theorem asserts that if f is analytic at every point in the complex plane, and if there is a number M such that $|f(z)| \le M$ for all $z \in \mathbb{C}$, then f must be a constant function. This is astonishing if one thinks of how *untrue* this is when the complex numbers are replaced by the real numbers. Consider, for example, the function $\sin x$. This function satisfies $|\sin x| \le 1$ for all $x \in \mathbb{R}$, and we can expand $\sin x$ as a Taylor series with real coefficients centered at any point that we please. Yet, $\sin x$ is certainly not a constant function.

We now consider a function f defined on an open subset U of the complex plane and taking values in a Banach space $(X, \| \cdot \|)$. This function is said to be *analytic* at $z_0 \in U$ if there is a positive number $r > 0$ and elements $a_1, a_2, \ldots$ in the Banach space such that $|z - z_0| < r$ implies $z \in U$ and

$$f(z) = \sum_{n=0}^{\infty} a_n (z - z_0)^n$$

for all $|z - z_0| < r$. The analogue of Liouville's theorem holds. Specifically, if f is a Banach space valued function that is analytic at every point of $\mathbb{C}$, and if there is a number M such that $\| f(z) \| \le M$ for all $z \in \mathbb{C}$, then f must be a constant function. A proof of the Banach space version of Liouville's theorem can be given using the Hahn–Banach theorem (see Section 6.3).

Theorem 5.7. *Let X be a Banach space. The spectrum of each element $T \in \mathcal{B}(X)$ is a compact and nonempty set, and is contained in the disk $\{\lambda \in \mathbb{C} \, \big| \, |\lambda| \le \|T\|\}$.*

PROOF.   We first prove that $\sigma(T)$ is a compact subset of $\mathbb{C}$; this part of the proof we can do in full. We begin by observing that the set $\mathbb{C} \setminus \sigma(T)$ is open. To see this, suppose that $\lambda \in \mathbb{C} \setminus \sigma(T)$ (so $(\lambda I - T)^{-1}$ exists) and that μ is a complex number satisfying

$$|\lambda - \mu| < \|(\lambda I - T)^{-1}\|^{-1}.$$

It follows from Theorem 5.6 that μ is also in $\mathbb{C} \setminus \sigma(T)$. Therefore, $\mathbb{C} \setminus \sigma(T)$ is open, and $\sigma(T)$ is consequently closed.

If $|\lambda| > \|T\|$, then $I - \frac{1}{\lambda}T$ is invertible by Theorem 5.5. Therefore,

$$\sigma(T) \subseteq \{\lambda \in \mathbb{C} \,\big|\, |\lambda| \le \|T\|\},$$

showing that $\sigma(T)$ is a bounded set. Since $\sigma(T)$ is closed and bounded, it is compact by the Heine–Borel theorem.

It remains to be shown that $\sigma(T)$ is nonempty. This is the part of the proof that requires Liouville's theorem and therefore is not to be considered complete. Define a function $f : \mathbb{C} \setminus \sigma(T) \to \mathcal{B}(X)$ by

$$f(\lambda) = (\lambda I - T)^{-1}.$$

This function is analytic (see, for example, [124], Theorem 2.3). If $\sigma(T)$ is assumed to be empty, then this function is analytic at each point of the complex plane. If $|\lambda| > \|T\|$, it follows from Theorem 5.5 that

$$f(\lambda) = (\lambda I - T)^{-1} = \frac{1}{\lambda}\left(I - \frac{1}{\lambda}T\right)^{-1} = \sum_{n=0}^{\infty} \frac{T^n}{\lambda^{n+1}}.$$

Summing a geometric series, we see that

$$\|f(\lambda)\| = \left\|\sum_{n=0}^{\infty} \frac{T^n}{\lambda^{n+1}}\right\| \le \sum_{n=0}^{\infty} \left\|\frac{T^n}{\lambda^{n+1}}\right\| \le \frac{1}{|\lambda| - \|T\|}.$$

This shows that

$$\|f(\lambda)\| \to 0$$

as $|\lambda| \to \infty$ and hence that $\|f(\lambda)\|$ is bounded. Liouville's theorem now implies that f must be constant. Since

$$\|f(\lambda)\| \to 0$$

as $|\lambda| \to \infty$, f must be identically zero. Since $f(\lambda)$ is defined to be $(\lambda I - T)^{-1}$, and it is impossible for an inverse to be zero, the spectrum of T must be nonempty, as desired. □

This proof is striking in that it applies a theorem about functions of a complex variable (Liouville's theorem) to prove a result about operators. Recall that to prove that a matrix actually has an eigenvalue one uses the fundamental theorem of algebra to assert that the characteristic polynomial of the matrix has a root. The fundamental theorem of algebra is not as easy to prove as it may appear, and one of the most elementary proofs uses Liouville's theorem! So, perhaps, the application of complex function theory used in the proof of the preceding theorem is not so surprising after all. Nonetheless, it is a beautiful proof in which analyticity and operator theory meet.

We point out that if $|\lambda| > \|T\|$, then $\lambda I - T$ is invertible, and its inverse is given by the series

$$\sum_{n=0}^{\infty} \frac{T^n}{\lambda^{n+1}}.$$

This series is called a "Laurent series." What if $\lambda I - T$ is invertible but $|\lambda| \leq \|T\|$? Is $(\lambda I - T)^{-1}$ still given by this Laurent series? The answer to this question is yes, but we are not in a position to understand a proof of this fact (see, for example, [124], Theorem 3.3). This same issue, of "extending" the representation of $(\lambda I - T)^{-1}$ as the Laurent series from the set $|\lambda| > \|T\|$ to the bigger set $\mathbb{C} \setminus \sigma(T)$, is exactly what prevents us from giving a complete proof of our next theorem (Theorem 5.8).

We define the *spectral radius* $r(T)$ of an element $T \in \mathcal{B}(X)$ to be

$$\sup\{|\lambda| \,\big|\, \lambda \in \sigma(T)\}.$$

One consequence of the preceding result is that

$$r(T) \leq \|T\|$$

for every bounded operator T. In the next section we will discover (in Theorem 5.13) a large class of operators for which this is an equality. Following that theorem we will give an example of an operator T satisfying $r(T) < \|T\|$.

The next theorem is remarkable in that it equates an apparently algebraic quantity (the spectral radius) with an analytic quantity. The result is often referred to as the *spectral radius formula*, and is extremely useful.

Theorem 5.8. *Let X be a Banach space and $T \in \mathcal{B}(X)$. Then*

$$r(T) = \lim_{n \to \infty} \|T^n\|^{\frac{1}{n}}.$$

PROOF.   The idea for this proof is to prove the two inequalities

$$r(T) \leq \liminf_{n \to \infty} \|T^n\|^{\frac{1}{n}},$$

and

$$\limsup_{n \to \infty} \|T^n\|^{\frac{1}{n}} \leq r(T).$$

For the first of these we can give a complete proof. The second requires, again, more analytic function theory than we have available. We now prove the parts that our background allows us to, and point out where more background knowledge of complex functions is needed.

Consider a complex number λ and positive integer n. Assume that $\lambda^n \notin \sigma(T^n)$, so that $(\lambda^n I - T^n)^{-1}$ exists. Notice that

$$\lambda^n I - T^n = (\lambda I - T)(\lambda^{n-1} I + \lambda^{n-2} T + \cdots + \lambda T^{n-2} + T^{n-1}),$$

and that the factors on the right commute. The operator

$$(\lambda^n I - T^n)^{-1}(\lambda^{n-1} I + \lambda^{n-2} T + \cdots + \lambda T^{n-2} + T^{n-1})$$

is seen to be the inverse of $\lambda I - T$. Therefore, $\lambda \notin \sigma(T)$. This shows that $\lambda \in \sigma(T)$ implies $\lambda^n \in \sigma(T^n)$ for each positive integer n. Theorem 5.7 now implies that

$$|\lambda^n| \leq \|T^n\|$$

and hence that

$$|\lambda| \leq \|T^n\|^{\frac{1}{n}}$$

for each positive integer n. The definition of spectral radius now gives

$$r(T) \leq \|T^n\|^{\frac{1}{n}}$$

for each positive integer n, and hence that

$$r(T) \leq \liminf_{n \to \infty} \|T^n\|^{\frac{1}{n}}.$$

This proves the first of the two inequalities.

We move on to the proof of

$$\limsup_{n \to \infty} \|T^n\|^{\frac{1}{n}} \leq r(T).$$

If $|\lambda| > \|T\|$, then the series

$$\sum_{n=0}^{\infty} \left\| \frac{T^n}{\lambda^{n+1}} \right\|$$

converges. This is a series of real numbers with radius of convergence

$$\frac{1}{|\lambda|} \limsup_{n \to \infty} \|T^n\|^{\frac{1}{n}}.$$

Since the series converges, it must be the case that

$$\frac{1}{|\lambda|} \limsup_{n \to \infty} \|T^n\|^{\frac{1}{n}} \leq 1,$$

or, equivalently,

$$\limsup_{n \to \infty} \|T^n\|^{\frac{1}{n}} \leq |\lambda|.$$

This holds for any $|\lambda| > \|T\|$; the proof would be completed by showing that this inequality holds for any $|\lambda| > r(T)$. Since, $r(T) \leq \|T\|$, we do have a chance, but this is where we cannot really go any further without knowing about "uniqueness of Laurent series" in complex analysis. The extension of the inequality

$$\limsup_{n \to \infty} \|T^n\|^{\frac{1}{n}} \leq |\lambda|$$

for all $|\lambda| > \|T\|$ to all $|\lambda| > r(T)$ follows immediately from the identity theorem of complex function theory. $\qquad\square$

The last two theorems are really at the edge of what we think we can cover in this text. They should give you an indication of how ideas from complex function theory can be used to prove operator-theoretic results. In spectral theory, much hinges on the fact the the operator $(\lambda I - T)^{-1}$ is given by a "geometric series" (as in Theorem 5.5).

One important class of operators is the class of "compact operators." Compact operators are, in some sense, the most natural generalization of finite-dimensional operators. The notion of a compact operator was first given by Hilbert, and their theory was greatly expanded by Riesz. Consider normed linear spaces X and Y. A linear operator $T : X \to Y$ is *compact* if for every bounded sequence $\{x_n\}_{n=1}^{\infty}$ in X, the sequence $\{Tx_n\}_{n=1}^{\infty}$ in Y has a convergent subsequence in Y. Compact operators are always bounded, as you are asked to prove in Exercise 5.3.11. Two fundamental properties of the collection of compact operators are given in the next two theorems.

Theorem 5.9. *Let X be a Banach space and $T \in \mathcal{B}(X)$ a compact operator. Then ST and TS are compact for each $S \in \mathcal{B}(X)$.*

PROOF.   Consider a bounded sequence $\{x_n\}_{n=1}^{\infty}$ in X. Since T is compact, the sequence $\{Tx_n\}_{n=1}^{\infty}$ has a convergent subsequence $\{Ty_n\}_{n=1}^{\infty}$ converging to y. Then

$$\|STy_n - Sy\| \leq \|S\| \cdot \|Ty_n - y\|,$$

showing that $\{STy_n\}_{n=1}^{\infty}$ converges to Sy. This shows that ST is compact. We note that the sequence $\{Sx_n\}_{n=1}^{\infty}$ is also bounded. Therefore, $\{TSx_n\}_{n=1}^{\infty}$ has a convergent subsequence, proving that TS is compact.   □

Theorem 5.10. *If X is a Banach space, then the set of compact operators from X to X is closed in $\mathcal{B}(X)$.*

PROOF.   Suppose that $T_n \in \mathcal{B}(X)$ is compact for each positive integer n, and that $\|T_n - T\| \to 0$ as $n \to \infty$. We aim to show that T is a compact operator.

Consider a bounded sequence $\{x_n\}_{n=1}^{\infty}$ in X, with

$$B = \sup\{|x_n| \,\big|\, 1 \leq n < \infty\}.$$

Let $\epsilon > 0$, and choose M large enough to satisfy

$$\|T_M - T\| < \frac{\epsilon}{3B}.$$

Since T_M is compact, there is a subsequence $\{y_n\}_{n=1}^{\infty}$ of $\{x_n\}_{n=1}^{\infty}$ such that $\{T_M y_n\}_{n=1}^{\infty}$ is convergent, and hence Cauchy. Choose N large enough so that

$$\|T_M y_n - T_M y_m\| < \frac{\epsilon}{3}$$

whenever $n, m \geq N$. We aim to show that $\{Ty_n\}_{n=1}^{\infty}$ is Cauchy. Then, since X is complete, it will follow that $\{Ty_n\}_{n=1}^{\infty}$ converges, as desired. As long as $n, m \geq N$, we have that

$$\|Ty_n - Ty_m\| \leq \|Ty_n - T_M y_n\| + \|T_M y_n - T_M y_m\| + \|T_M y_m - Ty_m\|$$
$$\leq \|T - T_M\| \cdot \|y_n\| + \frac{\epsilon}{3} + \|T_M - T\| \cdot \|y_m\|$$

$$\leq \frac{\epsilon}{3B} \cdot B + \frac{\epsilon}{3} + \frac{\epsilon}{3B} \cdot B = \epsilon,$$

proving that $\{Ty_n\}_{n=1}^{\infty}$ is Cauchy.   □

One reason that compact operators are relatively easy to work with, and hence so attractive, is because of the structure of their spectra. The spectrum of a compact operator is similar to the spectrum of a finite matrix. This information is gathered in the next theorem. It is, for the most part, due to F. Riesz, and is a masterpiece. It belongs to a collection of results commonly referred to as the Riesz theory for compact operators. We will not prove this result, but you may want to compare it to Theorem 5.23. See Exercise 5.3.18.

Theorem 5.11. *Let X be an infinite-dimensional Banach space and $T \in \mathcal{B}(X)$ a compact operator. Then the spectrum of T is either a finite set or is a sequence converging to 0. The point 0 is in the spectrum, and each nonzero value of the spectrum is an eigenvalue. Further, for $\lambda \neq 0$, the eigenspace $\ker(\lambda I - T)$ is finite-dimensional.*

We would like to have a practical way of deciding whether a given operator is compact. Are there relatively easy-to-use tests for deciding compactness? Our first example shows how Theorem 5.10 can be used to prove that a given operator is compact.

EXAMPLE 1.   The diagonal operator $D = \text{diag}(\lambda_1, \lambda_2, \ldots) \in \mathcal{B}(\ell^2)$ is compact if and only if $\lim_{n \to \infty} \lambda_n = 0$.

To see this, let $\{e_n\}_{n=1}^{\infty}$ denote the standard orthonormal basis of ℓ^2, and let

$$\overline{D} = \text{diag}(\overline{\lambda_1}, \overline{\lambda_2}, \ldots).$$

From Bessel's inequality we deduce that $\langle De_n, x \rangle = \langle e_n, \overline{D}x \rangle \to 0$ for every $x \in \ell^2$. Suppose that $|\lambda_n| = \|De_n\|_2$ does not converge to 0 as $n \to 0$. Then, there is a subsequence $\{f_n\}_{n=1}^{\infty}$ of $\{e_n\}_{n=1}^{\infty}$ such that $\|Df_n\|_2 \geq \epsilon$ for some $\epsilon > 0$, and every n. Since the sequence $\{f_n\}_{n=1}^{\infty}$ is bounded and D is compact, $\{f_n\}_{n=1}^{\infty}$ contains a subsequence $\{g_n\}_{n=1}^{\infty}$ such that $\{Dg_n\}_{n=1}^{\infty}$ converges in ℓ^2. Let y denote the element of ℓ^2 satisfying $\|Dg_n - y\|_2 \to 0$ as $n \to \infty$. Then

$$|\langle Dg_n - y, x \rangle| \leq \|Dg_n - y\|_2 \|x\|_2$$

for every $x \in \ell^2$, and hence $\langle Dg_n, x \rangle \to \langle y, x \rangle$ as $n \to \infty$. But $\langle Dg_n, x \rangle$ must converge to 0 by Bessel's inequality. Thus, $\langle y, x \rangle = 0$ for every $x \in \ell^2$, and hence $y = 0$, contradicting $\|Dg_n\|_2 \geq \epsilon$ for every n. Therefore, our supposition was incorrect, so that $\lim_{n \to \infty} |\lambda_n| = 0$, and hence $\lim_{n \to \infty} \lambda_n = 0$.

Conversely, suppose that $\lim_{n \to \infty} \lambda_n = 0$. Define the truncated diagonal operators

$$D_k = \text{diag}(\lambda_1, \lambda_2, \ldots, \lambda_k, 0, 0, \ldots).$$

We aim to show that each D_k is a compact operator. To this end, fix k, and consider a bounded sequence $\{x_n\}_{n=1}^{\infty}$ in ℓ^2. Each x_n is itself a sequence, and we let x_n^j

denote the jth entry of the sequence x_n. That is, for each n,

$$x_n = \{x_n^1, x_n^2, x_n^3, \ldots\}, \qquad \sum_{j=1}^{\infty} |x_n^j|^2 < \infty.$$

Let B denote a bound for the sequence $\{x_n\}_{n=1}^{\infty}$, so that

$$\sqrt{\sum_{j=1}^{\infty} |x_n^j|^2} = \|x_n\|_{\ell^2} \le B$$

for each $x_n \in \ell^2$. Recall that k is fixed, and note that

$$D_k x_n = \{\lambda_1 x_n^1, \lambda_2 x_n^2, \lambda_3 x_n^3, \ldots, \lambda_k x_n^k, 0, 0, \ldots\}.$$

If we arrange the images of $D_k x_1, D_k x_2, \ldots$ as rows in a matrix, we get

$$\begin{array}{cccc} \lambda_1 x_1^1 & \lambda_2 x_1^2 & \lambda_3 x_1^3 & \cdots \\ \lambda_1 x_2^1 & \lambda_2 x_2^2 & \lambda_3 x_2^3 & \cdots \\ \lambda_1 x_3^1 & \lambda_2 x_3^2 & \lambda_3 x_3^3 & \cdots \\ \vdots & \vdots & \vdots & \ddots \end{array}$$

Each column in this array is a bounded sequence of numbers; the jth column is bounded by, for example, $|\lambda_j|B$. Recall that bounded sequences always contain convergent subsequences. Each column, and in particular the *first* column, thus has a convergent subsequence, say

$$\{\lambda_1 x_{n_1}^1, \lambda_1 x_{n_2}^1, \lambda_1 x_{n_3}^1, \ldots\}.$$

The subsequence

$$\{\lambda_2 x_{n_1}^2, \lambda_2 x_{n_2}^2, \lambda_2 x_{n_3}^2, \ldots\}$$

of the second column, in turn, has a convergent subsequence. Continue in this way, until a convergent subsequence of the kth column has been produced. Abusing notation, denote this subsequence of the kth column by

$$\{\lambda_k x_1^k, \lambda_k x_2^k, \lambda_k x_3^k, \ldots\}.$$

Then, for each $j = 1, \ldots, k$, the subsequence

$$\{\lambda_j x_1^j, \lambda_j x_2^j, \lambda_j x_3^j, \ldots\}$$

of the jth column converges. For $\epsilon > 0$, we can thus choose N such that

$$|x_n^j - x_m^j| < \frac{\epsilon}{\sqrt{k}L_k}$$

for all $j = 1, \ldots, k$ and $n, m > N$, where $L_k = \max\{|\lambda_j| : 1 \le j \le k\}$. Then

$$\|D_k x_n - D_k x_m\|_{\ell^2}^2 = \sum_{j=1}^{k} |\lambda_j|^2 \cdot |x_n^j - x_m^j|^2 \le \sum_{j=1}^{k} L_k^2 \left(\frac{\epsilon}{\sqrt{k}L_k}\right)^2 < \epsilon^2.$$

This shows that the sequence $\{D_k x_n\}_{n=1}^{\infty}$ is Cauchy, and hence converges. This completes the proof that the operator D_k is compact.

Next, our hypothesis that $\lim_{n \to \infty} \lambda_n = 0$ implies that

$$\|D_k - D\| = \sup\{|\lambda_{k+i}| \,|\, i = 0, 1, 2, \ldots\} \to 0.$$

Theorem 5.10 now implies that D is compact.

The operators D_k are examples of *finite-rank* operators: operators with finite-dimensional range. All finite rank operators are compact. To show that a given operator is compact, it suffices to see that it is the limit of finite-rank operators, and this is a standard way of showing that a given operator is, in fact, compact. It is *not* the case that every compact operator is the limit of finite rank operators. However, on most "nice" Banach spaces this will be the case (for example, on all Hilbert spaces). The first counterexample to this so-called approximation problem was published in 1973 by the Swedish mathematician Per Enflo (born 1944). His example also gave a negative solution to the so-called basis problem. Both problems just mentioned were long-standing important open problems in analysis. There is a third famous problem that Enflo is responsible for resolving, and it will be the theme for the next section.

EXAMPLE 2.   The left shift $T \in \mathcal{B}(\ell^2)$ is not compact. The sequence $\{x_n\}_{n=1}^{\infty} \in \ell^2$, where x_n is the ℓ^2 element $(0, \ldots, 0, 1, 0, 0, \ldots)$ (the 1 appears in the nth place) is a bounded sequence, with $\|x_n\|_{\ell^2} = 1$ for each n. However, the sequence $\{Tx_n\}_{n=1}^{\infty}$ has no convergent subsequence. If it did, this subsequence would have to be Cauchy. This is impossible, since $n \neq m$ implies $\|Tx_n - Tx_m\|_{\ell^2} = 2$.

EXAMPLE 3.   The identity operator I on an infinite-dimensional Banach space X cannot be compact. This follows from Exercise 2.1.13(c).

We conclude this long section with the computation of the spectra of a few operators. Computing spectra can be a very difficult exercise! Let $\text{Eig}(T)$ denote the set of all eigenvalues of T. Note that it is always the case that $\text{Eig}(T) \subseteq \sigma(T)$, but the larger set might be much larger.

EXAMPLE 4.   Let $\phi : [0, 1] \to \mathbb{R}$ be continuous and define a multiplication operator M_ϕ on $L^2([0, 1])$ by

$$M_\phi f(x) = \phi(x) f(x)$$

for $t \in [0, 1]$ and $f \in L^2([0, 1])$. In Exercise 5.3.13 you are asked to show that $M_\phi \in \mathcal{B}\left(L^2([0, 1]), \mathbb{R}\right)$. We claim that

$$\sigma(M_\phi) = f([0, 1]).$$

If $\lambda \notin f([0, 1])$, then (as you are asked to show in the same exercise) the multiplication operator $M_{(\lambda-\phi)^{-1}}$ is a bounded operator and is the inverse of $\lambda I - M_\phi$. Now suppose that $\lambda \in f([0, 1])$ and that $\lambda I - M_\phi$ has an inverse, T, in $\mathcal{B}\left(L^2([0, 1])\right)$. Since $\lambda \in f([0, 1])$, there is an $x_0 \in [0, 1]$ with $f(x_0) = \lambda$. Since f is continuous, we can pick, for each positive integer n, a number $\delta_n > 0$ such that $|f(x) - \lambda| < \frac{1}{n}$ for each x in the interval $(x_0 - \frac{\delta_n}{2}, x_0 + \frac{\delta_n}{2})$. Define a function g_n on $[0, 1]$ to take the value $\delta_n^{-\frac{1}{2}}$ on the interval $(x_0 - \frac{\delta_n}{2}, x_0 + \frac{\delta_n}{2})$ and 0 off of this interval. Note that $\|g_n\| = 1$ for each n. Put $h_n = (\lambda I - M_\phi)g_n$. Then $h_n \to 0$ as $n \to \infty$, yet $Th_n = g_n$ and hence $\{Th_n\}$ cannot converge to 0 as $n \to \infty$, showing that T is not continuous, and hence not bounded.

EXAMPLE 5.   Consider the left shift T on $\ell^2(\mathbb{N})$ given by the formula

$$T(x_1, x_2, \ldots) = (x_2, x_3, \ldots).$$

As we have seen in Section 2, $\|T\| = 1$. Therefore, by Theorem 5.7,

$$\sigma(T) \subseteq \{\lambda \in \mathbb{C} \mid |\lambda| \leq 1\}.$$

For $0 < |\lambda| < 1$, put $x_n = \lambda^{n-1}$. Then $\{x_n\}_{n=1}^\infty \in \ell^2(\mathbb{N})$ and

$$T(x_1, x_2, \ldots) = (\lambda, \lambda^2, \lambda^3, \ldots) = \lambda(x_1, x_2, \ldots),$$

showing that λ is an eigenvalue for T. If $\lambda = 0$, then $(1, 0, 0, \ldots)$ is an eigenvector. Thus $\sigma(T) \supseteq \{\lambda \in \mathbb{C} \mid |\lambda| < 1\}$. Since $\sigma(T)$ must be closed, we see that $\sigma(T)$ must, in fact, be exactly the closed unit disk. "Most" of the spectral values of the left shift are eigenvalues.

Now consider the right shift S given by

$$S(x_1, x_2, \ldots) = (0, x_1, x_2, \ldots).$$

We know that $\sigma(S)$ is not empty, so there must exist some $\lambda \in \sigma(S)$. If this λ were an eigenvalue then

$$(0, x_1, x_2, \ldots) = S(x_1, x_2, \ldots) = \lambda(x_1, x_2, \ldots)$$

for some nonzero element $(x_1, x_2, \ldots) \in \ell^2$. This is impossible. So, while the left shift has many eigenvalues, the right shift has none.

Since $\|S\| = 1$, we know that $\sigma(S)$ is contained in the closed unit disk. It turns out that $\sigma(S) = \sigma_{B(\ell^2(\mathbb{N}))}(S)$ is also the closed unit disk $\{\lambda \in \mathbb{C} : |\lambda| \leq 1\}$, but this is not as easy to prove as it was for T. See page 45 of [58] for a proof.

EXAMPLE 6.   The weighted shift on $\ell^2(\mathbb{N})$, given by the formula

$$W(x_1, x_2, \ldots) = \left(0, x_1, \frac{1}{2}x_2, , \frac{1}{3}x_3, \ldots\right),$$

is compact and has no eigenvalues. Why is this the case? This operator has matrix

$$\begin{pmatrix} 0 & 0 & 0 & 0 & \ldots \\ 1 & 0 & 0 & 0 & \ldots \\ 0 & \dfrac{1}{2} & 0 & 0 & \ldots \\ 0 & 0 & \dfrac{1}{3} & 0 & \ldots \\ \vdots & \vdots & \vdots & \vdots & \ddots \end{pmatrix}.$$

To see that W is a compact operator, note that $W = SD$, where S is the usual right shift and D is the diagonal operator $D = \mathrm{diag}\left(1, \frac{1}{2}, \frac{1}{3}, \ldots\right)$. As in Example 1, the diagonal operators $D_n = \text{diag}(1, \cdots, \frac{1}{n}, 0, 0, \cdots)$ are compact, and

$$\|D - D_n\| \leq \frac{1}{n+1} \to 0$$

as $n \to \infty$. We thus apply Theorem 5.10 to conclude that D is compact, and now the compactness of W follows from Theorem 5.9. In Exercise 5.3.14 you are asked to show that W has no eigenvalues. One could now use Theorem 5.11 to conclude that the spectrum of W must equal the singleton $\{0\}$. This argument is a bit of a cheat, since we have not proved Theorem 5.11. Alternatively, one can show that $\|W^n\| \leq \frac{1}{n!}$ for each positive integer n (see Exercise 5.3.14). Therefore, we see that

$$\|W^n\|^{\frac{1}{n}} \leq \left(\frac{1}{n!}\right)^{\frac{1}{n}} \to 0$$

as $n \to \infty$, and we can now apply Theorem 5.8 (note that we do not need the full statement of this theorem; we need only the inequality $r(W) \leq \liminf_{n \to \infty} \|W^n\|^{\frac{1}{n}}$). It is also possible to write down, explicitly, the inverse of the infinite matrix $\lambda I - W$ for any nonzero λ, thus showing that $\sigma(W) = \{0\}$. It is not a bad idea to do this more "elementary" proof. We gave the proof we did to demonstrate the use of the spectral radius formula.

In this example, the sequence $\{\frac{1}{n}\}_{n=1}^{\infty}$ can be replaced by an arbitrary sequence of positive numbers that decrease to zero.

An operator $T \in \mathcal{B}(X)$ is called *quasinilpotent* if $\sigma(T) = \{0\}$. By Theorem 5.8, this is equivalent to $r(T) = \{0\}$. The weighted shift W of Example 3 is quasinilpotent. See Exercise 5.3.12 for another example of a quasinilpotent operator.

A compact operator with no eigenvalues must be quasinilpotent (by Theorem 5.11). Compact and quasinilpotent operators will be revisited in the next section.

5.4   An Introduction to the Invariant Subspace Problem

Given a bounded linear operator T on a Banach space X, a subspace Y of X is called an *invariant subspace* for T if $T(Y) \subseteq Y$. The trivial subspaces $\{0\}$ and X are invariant for any $T \in \mathcal{B}(X)$ and any Banach space X. It is still not known whether there is a bounded linear operator on a Hilbert space that has only the trivial invariant subspaces. This is the *invariant subspace problem*. There are, however, examples of operators on Banach spaces with no nontrivial invariant subspace. The first such example was given by Per Enflo in 1987 [40]. There are now other examples, including examples due to Charles Read for which the underlying space X is the well-known sequence space ℓ^1. Read's examples appeared in papers published in the mid 1980s. For a good expository account of progress on the invariant subspace problem through the mid 1980s, read the introduction of [15]. This reference also contains one of Read's examples, as well as an extensive bibliography on the subject. For a most enjoyable introduction to the subject, [101] is highly recommended.

Even though it is known that there are operators with no nontrivial invariant subspaces, matters are not settled. Interesting positive results have been achieved by studying certain classes of operators. That is, there are certain Banach spaces X and certain types of bounded linear operators T on X for which it is known that there will always be a nontrivial invariant subspace. Theorems 5.12, 5.21, and 5.22 of this section give examples of these positive results.

We start with a consideration of the invariant subspaces of the right shift on $\ell^2 = \ell^2(\mathbb{N})$. Let

$$S(x_1, x_2, \ldots) = (0, x_1, x_2, \ldots)$$

and

$$M_n = \{(x_1, x_2, \ldots) \in \ell^2 : x_k = 0, 1 \le k \le n\}.$$

While it is straightforward to check that M_n is an invariant subspace for S, for each positive integer n, the answer to the question "are there any other invariant subspaces?" is not so obvious. See page 83 of [58] for a characterization of the invariant subspaces of the right shift on ℓ^2.

Explicit descriptions of the invariant subspaces of an operator are not so easy to come by, and we are usually happy just to know that invariant subspaces exist.

Theorem 5.12. *Let X be a finite-dimensional complex Banach space. Then every $T \in \mathcal{B}(X)$ has a nontrivial invariant subspace.*

PROOF. Suppose $\dim(X) = n < \infty$. Choose any nonzero vector $x \in X$. Then the set $\{x, Tx, \ldots, T^n x\}$ is linearly dependent because it contains $n + 1$ elements. Therefore, there exist scalars $\alpha_0, \alpha_1, \ldots, \alpha_n$, not all zero, such that

$$0 = \alpha_0 x + \alpha_1 Tx + \cdots + \alpha_n T^n x.$$

The complex polynomial $\alpha_0 + \alpha_1 z + \cdots + \alpha_n z^n$ can be factored as

$$\alpha(z - \lambda_1) \cdots (z - \lambda_m),$$

for some scalars $\alpha, \lambda_1, \ldots, \lambda_m$. Then

$$0 = \alpha_0 x + \alpha_1 Tx + \cdots + \alpha_n T^n x = (\alpha_0 + \alpha_1 T + \cdots + \alpha_n T^n)x$$
$$= \alpha(T - \lambda_1 I) \cdots (T - \lambda_m I)x,$$

where I is the $n \times n$ identity matrix. It follows that $T - \lambda_j I$ has a nonzero kernel for at least one value j. The corresponding eigenspace $\ker(T - \lambda_j I)$ is an invariant subspace for T. If T is not a multiple of the identity, $\ker(T - \lambda_j I)$ is properly contained in X and the theorem is proved. If T is a multiple of the identity then every subspace is invariant, and again the theorem is proved.   □

The proof of Theorem 5.12 is taken from [8], Theorem 2.1. Observe that if $T^m x = 0$ for some $m < n$ (as is the case, for example, for any upper-triangular matrix with zeroes on the diagonal), then the closed subspace spanned by the vectors $\{x, Tx, \ldots, T^{m-1}x\}$ is also a nontrivial invariant subspace for T.

Next, we consider operators on a complex Hilbert space H. Hilbert spaces have more structure than general normed linear spaces. In fact, they have so much structure that in many cases problems that are intractable on general normed linear spaces become trivial on Hilbert spaces. In the case of the invariant subspace problem, they have enough structure so that we can prove some interesting positive results, yet the problem remains unsolved.

An operator $T \in \mathcal{B}(H)$ is *Hermitian* if

$$\langle Tx, y \rangle = \langle x, Ty \rangle$$

for all $x, y \in H$. These operators are named in honor of Charles Hermite. Recall, by Exercise 5.3.5, that $\|T^n\| \leq \|T\|^n$ for any linear operator (Hermitian or not) and any positive integer n. If T is now assumed Hermitian, then the Cauchy–Schwarz inequality and definition of operator norm imply that

$$\|Tx\|^2 = \langle Tx, Tx \rangle = \langle T^2 x, x \rangle \leq \|T^2\| \cdot \|x\|^2.$$

For $x \neq 0$ this implies that

$$\frac{\|Tx\|^2}{\|x\|^2} \leq \|T^2\|,$$

and taking the supremum over all x of norm 1 yields

$$\|T\|^2 \leq \|T^2\|.$$

Induction can now be used to obtain

$$\|T\|^{2^m} \leq \|T^{2^m}\|$$

for each positive integer m. Hence

$$\|T^{2^m}\| = \|T\|^{2^m}$$

for each positive integer m. Let $1 \leq n \leq 2^m$. Then

$$\begin{aligned}
\|T^{2^m}\| &= \|T^n T^{2^m - n}\| \\
&\leq \|T^n\| \cdot \|T^{2^m - n}\| \\
&\leq \|T^n\| \cdot \|T\|^{2^m - n} \\
&\leq \|T\|^n \cdot \|T\|^{2^m - n} \\
&= \|T\|^{2^m},
\end{aligned}$$

so that all of these expressions are, in fact, equal. In particular,

$$\|T^n\| \cdot \|T\|^{2^m - n} = \|T\|^{2^m},$$

so that

$$\|T^n\| \cdot \|T\|^{-n} = 1.$$

We have now proved that for a Hermitian operator T and a positive integer n,

$$\|T^n\| = \|T\|^n.$$

Theorem 5.13. *Let H be a Hilbert space. If $T \in \mathcal{B}(H)$ is Hermitian, then* $r(T) = \|T\|$.

PROOF.   The proof follows immediately from the equality $\|T^n\| = \|T\|^n$ and the spectral radius formula (Theorem 5.8):

$$r(T) = \lim_{n \to \infty} \|T^n\|^{\frac{1}{n}} = \lim_{n \to \infty} \|T\| = \|T\|. \qquad \square$$

If we consider the operator

$$T = \begin{pmatrix} 1 & 1 \\ 0 & 1 \end{pmatrix} \in \mathcal{B}(\ell^2),$$

then $r(T) = \max\{|\lambda_1|, |\lambda_1|\}$, where λ_1, λ_2 are the eigenvalues of T, and thus $r(T) = 1$. However, $\|T\| = \sqrt{\frac{1}{2}(3 + \sqrt{5})}$, as you are asked to compute in Exercise 5.2.6. This provides an example with $r(T) < \|T\|$.

Our next immediate goal is to prove that each spectral value of a Hermitian operator is a real number. Compare this to what you know about the eigenvalues of a Hermitian matrix (see Exercise 5.4.1). In order to show this we make use of the notion of an "orthogonal complement."

For a subspace K of a Hilbert space H we define its *orthogonal complement* $K^\perp$ to be

$$K^\perp = \{y \in H \,|\, \langle x, y \rangle = 0 \text{ for all } x \in K\}.$$

The next theorem records two basic properties of orthogonal complements.

Theorem 5.14. *Suppose that K is a subspace of a Hilbert space H.*

(a) $K^\perp$ *is a closed subspace of H.*
(b) $K^\perp \cap K = \{0\}$.

PROOF.   To prove (a) assume that the sequence $\{y_n\}_{n=1}^{\infty}$ converges to y in H and that for each n, $y_n \in K^\perp$. Then, for each $x \in K$,

$$|\langle y, x \rangle| \le |\langle y - y_n, x \rangle| + |\langle y_n, x \rangle| \le \|y - y_n\| \cdot \|x\| + 0,$$

which can be made as small as we wish, and therefore $|\langle y, x \rangle| = 0$, as desired.

The proof of (b), as you should check, is completely straightforward. $\qquad \square$

The next theorem is interesting and has, as a consequence, many useful corollaries. Included among these corollaries is the important Riesz–Fréchet theorem characterizing the dual space of a Hilbert space. The definition of the "dual space" of any normed linear space can be found in Section 6.3. For a proof of the Riesz–Fréchet theorem see, for example, [129].

Theorem 5.15. *Assume that K is a closed subspace of a Hilbert space H and that $x \in H$. Then there exists $y \in K$ such that*

$$\|x - y\| = \inf\{\|x - z\| \,|\, z \in K\}.$$

Further, the element y is the unique element of K with this property.

PROOF.   As a convenience, let d denote $\inf\{\|x - z\| : z \in K\}$. Then there exists a sequence $\{y_n\}_{n=1}^{\infty}$ in K such that $\|x - y_n\| \to d$ as $n \to 0$. Observe that $\frac{1}{2}(y_n + y_m) \in K$ and thus

$$d \le \left\| x - \frac{y_n + y_m}{2} \right\| \le \frac{1}{2}\|x - y_n\| + \frac{1}{2}\|x - y_m\|.$$

This shows that

$$\left\| x - \frac{y_n + y_m}{2} \right\| \to d$$

as $n, m \to \infty$. In the computation that follows we use the parallelogram equality. Since

$$\|y_n - y_m\|^2 = \|(x - y_m) + (y_n - x)\|^2$$
$$= 2(\|x - y_m\|^2 + \|y_n - x\|^2) - \|2x - (y_n + y_m)\|^2$$
$$= 2(\|x - y_m\|^2 + \|y_n - x\|^2) - 4\left\| x - \frac{y_n + y_m}{2} \right\|^2$$

and all three normed terms in the last expression converge to d, we see that $\{y_n\}_{n=1}^{\infty}$ is a Cauchy sequence. Since K is closed, it is complete, and so K contains the limit point, y, of this sequence. Since $y \in K$, it follows that $\|x - y\| \ge d$. Since we assumed that $\|x - y_n\| \to d$, we could have chosen $\{y_n\}_{n=1}^{\infty}$ to satisfy, for example,

$$\|x - y_n\| \le d + \frac{1}{n}.$$

Letting $n \to \infty$ then yields $\|x - y\| \le d$, and thus $\|x - y\| = d$, as desired.

The final assertion of the theorem is left as Exercise 5.4.4.   □

Theorem 5.16. *Assume that K is a closed subspace of a Hilbert space H and that $x \in H$. Then x can be written as a sum $y + z$ for some $y \in K$ and $z \in K^{\perp}$. Moreover, this decomposition is unique.*

PROOF.   Let y be the unique point in K that is closest to x. Such a point is guaranteed by the preceding theorem. Put $z = x - y$. We will be done when we show that $z \in K^{\perp}$. For any $w \in K$, $y + w \in K$, and thus, by definition of y,

$$\|z\| = \|x - y\| \le \|x - (y + w)\| = \|(x - y) - w\|.$$

So we have

$$\|z - w\| \ge \|z\|$$

for each $w \in K$. Then, for $\lambda \in \mathbb{C}$, we have $\lambda w \in K$, and so

$$\|z - \lambda w\| \ge \|z\|, \qquad \text{or} \qquad \|z - \lambda w\|^2 \ge \|z\|^2.$$

Expanding the left side of this we get

$$\|z\|^2 \le \|z - \lambda w\|^2$$
$$= \langle z - \lambda w, z - \lambda w \rangle$$
$$= \langle z, z \rangle - \overline{\lambda}\langle z, w \rangle - \overline{\lambda}\langle z, w \rangle + |\lambda|^2 \langle w, w \rangle$$

$$= \|z\|^2 - 2 \cdot \mathrm{re}\Big\{\bar{\lambda}\langle z, w\rangle\Big\} + |\lambda|^2 \|w\|^2.$$

This holds for all $\lambda \in \mathbb{C}$, and in particular for $\lambda = re^{i\theta}$ where $r > 0$ and θ is chosen to satisfy $e^{-i\theta}\langle z, w\rangle = |\langle z, w\rangle|$. Using this choice of λ yields

$$-2r|\langle z, w\rangle| + r^2 \|w\|^2 \geq 0,$$

or

$$|\langle z, w\rangle| \leq \frac{1}{2}r\|w\|^2.$$

Since this holds for each $r > 0$, we have $\langle z, w\rangle = 0$. Since this holds for each $w \in K$, we have that $z \in K^\perp$, as desired.

To see that the decomposition is unique, suppose that $x = y_0 + z_0$ for $y_0 \in K$ and $z_0 \in K^\perp$. We will show that $y_0 = y$. For any $w \in K$, we have $\langle x - y_0, y_0 - w\rangle = 0$, and so

$$\|x - w\|^2 = \|x - y_0 + y_0 - w\|^2 = \|x - y_0\|^2 + \|y_0 - w\|^2 \geq \|x - y_0\|^2.$$

Therefore, $d = \|x - y_0\|$. By the uniqueness in Theorem 5.15, $y_0 = y$. □

From this theorem one can deduce that $(K^\perp)^\perp = K$ for a closed subspace K of a Hilbert space H (see Exercise 5.4.3). Observe also that $\{0\}^\perp = H$ and $H^\perp = \{0\}$.

We now return to our study of Hermitian operators.

Theorem 5.17. *Let H be a Hilbert space. If $T \in \mathcal{B}(H)$ is Hermitian, then $\ker(T)$ and $\mathrm{range}(T)$ are subspaces of H and satisfy*

$$\ker(T)^\perp = \mathrm{range}(T) \qquad \text{and} \qquad \mathrm{range}(T)^\perp = \ker(T).$$

($\ker(T)$ *and* $\mathrm{range}(T)$ *denote, respectively, the kernel and range of the map T).*

PROOF.   Left as Exercise 5.4.5.   □

Theorem 5.18. *Let H be a Hilbert space. If $T \in \mathcal{B}(H)$ is Hermitian, then $\sigma(T) \subseteq \mathbb{R}$.*

PROOF.   Consider $\lambda = a + ib \in \mathbb{C}$. Then, for each $x \in H$,

$$\|(\lambda I - T)x\|^2 = \Big\langle (\lambda I - T)x, (\lambda I - T)x \Big\rangle$$
$$= \Big\langle (aI - T)x, (aI - T)x \Big\rangle + \langle ibx, ibx\rangle$$
$$= \|(aI - T)x\|^2 + b^2\|x\|^2$$
$$\geq b^2\|x\|^2.$$

Therefore,

$$\|(\lambda I - T)x\| \geq |b| \cdot \|x\|,$$

and so if $b \neq 0$, then $\ker(\lambda I - T) = \{0\}$. The preceding theorem now gives $\mathrm{range}(\lambda I - T) = H$. So now we have shown that whenever $b \neq 0$, the operator $\lambda I - T$ is one-to-one and onto. Define S on H by $Sx = y$ precisely when

$(\lambda I - T)y = x$. We aim to show that S is continuous on H. This S will then be the inverse for $\lambda I - T$ in $\mathcal{B}(H)$, and thus we will have shown that $\lambda I - T$ is invertible whenever $b \neq 0$. From this it follows, by definition, that $\sigma(T) \subseteq \mathbb{R}$, as desired.

To see that S is continuous it suffices to see that S is continuous at 0 (see the comments following Theorem 5.1). Consider a sequence $\{x_n\}_{n=1}^{\infty}$ converging to 0 in H. Then, since $\lambda I - T$ is onto, there exists a sequence of points $\{y_n\}_{n=1}^{\infty}$ in H satisfying $x_n = (\lambda I - T)y_n$ for each n. Since $(\lambda I - T)y_n \to 0$ and

$$\|(\lambda I - T)y_n\| \geq |b| \cdot \|y_n\|,$$

we see that $Sx_n = y_n \to 0$, completing the proof.   □

Recall that the eigenvalues of a linear operator T form a (sometimes empty) subset of the spectrum of T. Observe that Theorem 5.13, the preceding theorem, and the fact that the spectrum must be a closed set show that at least one of $-\|T\|$ and $\|T\|$ is in the spectrum of a Hermitian operator T. Our next theorem gives even more precise information about this special spectral value in case T is also a compact operator. Its proof depends on the observation made in the following lemma, which is interesting in its own right.

Lemma 5.19. *Let H be a Hilbert space. If $T \in \mathcal{B}(H)$ is Hermitian, then*

$$\|T\| = \sup\{|\langle Tx, x \rangle| \,\big|\, \|x\| = 1\}.$$

PROOF.   For convenience, let M denote $\sup\{|\langle Tx, x \rangle| \,\big|\, \|x\| = 1\}$. If $\|x\| = 1$, then

$$|\langle Tx, x \rangle| \leq \|Tx\| \cdot \|x\| \leq \|T\| \cdot \|x\|^2 = \|T\|.$$

Since this holds for every x of norm 1, we see that $M \leq \|T\|$. To see the other inequality consider x of norm 1 and set $y = \frac{Tx}{\|Tx\|}$. Using the parallelogram equality we deduce that

$$
\begin{aligned}
\|Tx\| &= \langle Tx, y \rangle \\
&= \frac{1}{4}\Big[\big\langle T(x+y), x+y \big\rangle - \big\langle T(x-y), x-y \big\rangle\Big] \\
&\leq \frac{1}{4}M\Big[\|x+y\|^2 - \|x-y\|^2\Big] \\
&= \frac{1}{2}M\Big[\|x\|^2 + \|y\|^2\Big] \\
&= M.
\end{aligned}
$$

By definition of the operator norm, $\|T\| \leq M$, as desired.   □

Theorem 5.20. *Let H be a Hilbert space. If $T \in \mathcal{B}(H)$ is compact and Hermitian, then at least one of $-\|T\|$ and $\|T\|$ is an eigenvalue of T.*

PROOF.   By the preceding lemma we can find a sequence $\{x_n\}_{n=1}^{\infty}$ in H satisfying $\|x_n\| = 1$ and $|\langle Tx_n, x_n \rangle| \to \|T\|$. Since

$$\langle Tx_n, x_n \rangle = \langle x_n, Tx_n \rangle = \overline{\langle Tx_n, x_n \rangle},$$

we see that each $\langle Tx_n, x_n \rangle$ is, in fact, real. Depending on whether

$$\sup\{\langle Tx, x \rangle \mid \|x\| = 1\} \qquad \text{or} \qquad \inf\{\langle Tx, x \rangle \mid \|x\| = 1\}$$

has the larger absolute value, the sequence $\{\langle Tx_n, x_n \rangle\}_{n=1}^{\infty}$ converges to either $\|T\|$ or $-\|T\|$. We assume that $\{\langle Tx_n, x_n \rangle\}_{n=1}^{\infty}$ converges to $\|T\|$. The proof in the other case is identical. We will show that $\|T\|$ is an eigenvalue of T. Since T is compact, there is a subsequence $\{y_n\}_{n=1}^{\infty}$ of $\{x_n\}_{n=1}^{\infty}$ such that $\{Ty_n\}_{n=1}^{\infty}$ converges. Let y denote the limit of this sequence. Notice that as $n \to \infty$,

$$\left\| Tx_n - \|T\|x_n \right\|^2 = \|Tx_n\|^2 - 2\|T\|\langle Tx_n, x_n \rangle + \|T\|^2\|x_n\|^2$$

$$\leq 2\|T\|^2 - 2\|T\|\langle Tx_n, x_n \rangle \to 0.$$

Therefore, as $n \to \infty$,

$$\|T\|y_n = \left(\|T\|y_n - Ty_n \right) + Ty_n \to 0 + y = y.$$

Applying T to this, we get

$$\|T\|Ty_n = T\left(\|T\|y_n \right) \to Ty.$$

But we also know that $\|T\|Ty_n \to \|T\|y$, telling us that

$$Ty = \|T\|y.$$

As long as $y \neq 0$, we have found that $\|T\|$ is an eigenvalue of T. Can $y = 0$? Since

$$\|y\| = \lim_{n \to \infty} \|Ty_n\| = \lim_{n \to \infty} \left\| \|T\|y_n + Ty_n - \|T\|y_n \right\|$$

$$\geq \lim_{n \to \infty} \left(\left\| \|T\|y_n \right\| - \left\| Ty_n - \|T\|y_n \right\| \right)$$

$$= \lim_{n \to \infty} \left(\|T\| - \left\| Ty_n - \|T\|y_n \right\| \right)$$

$$= \|T\| > 0,$$

y cannot possibly be 0, and the proof is complete.   □

We now return to our study of invariant subspaces, and end the section with two positive results proving existence of invariant subspaces for certain classes of operators. In terms of trying to get information about invariant subspaces, notice that $\ker(\lambda I - T)$ is invariant under T for every complex number λ.

Theorem 5.21. *Every compact Hermitian operator on an infinite-dimensional Hilbert space has an invariant subspace.*

PROOF.   By our work above, such an operator T has an eigenvalue λ. The kernel of $\lambda I - T$ is thus a nonzero subspace. Since H is infinite-dimensional and T is compact, T cannot equal λI, and so this kernel is a proper subspace. Since T commutes with $\lambda I - T$, this subspace is invariant under the action of T.   □

What if we drop the hypothesis that T is Hermitian? Do we still get an invariant subspace? In particular, we might be interested in T acting on a Banach space (not necessarily a Hilbert space). Because the kernel of the operator $\lambda I - T$ is always an invariant subspace, we know that *if* a compact operator has an eigenvalue, then this kernel is nonzero and we get an invariant subspace. Theorem 5.11 makes this look promising. However, not all compact operators have an eigenvalue. As we saw in Example 6 of the preceding section, the weighted shift

$$W(x_1, x_2, \ldots) = (0, w_1 x_1, w_2 x_2, \ldots), \qquad w_n \downarrow 0,$$

on $\ell^2(\mathbb{N})$ is compact and has no eigenvalues. Nonetheless, it is true that every compact operator has a nontrivial invariant subspace. This result was proved by von Neumann in the 1930s, but was not published. The first published proof appeared in 1954 in [5].

In 1973, a theorem was proved that subsumes a proof that every compact operator has a nontrivial invariant subspace. It is due to Victor Lomonosov [84]. Lomonosov's result is remarkable, and it immediately drew much attention. The history of the events leading up to Lomonosov's work, and an account of the repercussions this work has had, are beautifully told in [101]. Interestingly, [101] was written after Enflo's famous paper (mentioned in the opening paragraph of this section) was put into circulation in the mathematical community, but before its publication in 1987. It offers information on why it took so long to publish Enflo's article. It also outlines Enflo's approach to his solution of the invariant subspace problem.

Very shortly after Lomonosov's proof began circulating in the community, Hugh Hilden gave another proof. We now state a version of Lomonosov's theorem, and give Hilden's proof of it.

Theorem 5.22. *Assume that X is an infinite-dimensional (complex) Banach space and that $T \in \mathcal{B}(X)$ satisfies $TK = KT$ for some nonzero compact operator $K \in \mathcal{B}(X)$. Then T has a nontrivial invariant subspace.*

PROOF.   Assume that T has no nontrivial invariant subspace. If K has an eigenvalue λ, then the kernel of the operator $\lambda I - K$ is seen to be an invariant subspace for T. Since K is compact, and X is infinite-dimensional, this kernel is a nontrivial subspace of X. Therefore, K must not have an eigenvalue. By Theorem 5.11, $\sigma(K) = \{0\}$. By the spectral radius formula,

$$\lim_{n \to \infty} \|(\alpha K)^n\|^{\frac{1}{n}} = 0$$

for every complex number α.

We may assume that $\|K\| = 1$ (if it does not, use $\frac{K}{\|K\|}$ in place of K). Choose an $x_0 \in X$ such that $\|Kx_0\| > 1$. Observe that $\|x_0\|$ must be greater than 1. Consider the closed ball

$$B = \{x \in X \mid \|x - x_0\| \le 1\},$$

and notice that $0 \notin B$. For any $x \neq 0$ in X, the closure of the set

$$\{p(T)x \mid p \text{ is a complex polynomial}\}$$

is a closed nonzero invariant subspace for T. By hypothesis, every closed nonzero invariant subspace for T is all of X. For a fixed $x \neq 0$ in X and any open set U in X, there must thus be a polynomial p such that $p(T)x \in U$.

Define

$$U_p = \{x \in X \mid \|p(T)x - x_0\| < 1\}.$$

Each U_p is open in X, and every nonzero element of X is in at least one of the U_p's. By Exercise 5.3.2, $K(B)$ has compact closure. Since the U_p's form an open cover of $K(B)$, there exist polynomials $p_1, \ldots, p_N$ such that

$$K(B) \subseteq \bigcup_{k=1}^{N} U_{p_k}.$$

In particular, if $x \in K(B)$, then there is a polynomial p_k for some $1 \leq k \leq N$ such that $p_k(T)x \in B$.

Up until this point in the proof, Hilden follows Lomonosov. At this point, their methods diverge. Lomonosov makes use of the Schauder fixed point theorem. We now give the rest of Hilden's proof.

Since $K x_0 \in K(B)$, $p_{k_1}(T)K x_0 \in B$ for some k_1. Then $K p_{k_1}(T)K x_0 \in K(B)$, and so $p_{k_2}(T)K p_{k_1}(T)K x_0 \in B$ for some k_2. Continue this process; after n steps, we get

$$p_{k_m}(T)K \cdots p_{k_2}(T)K p_{k_1}(T)K x_0 \in B,$$

for some $k_1, k_2, \ldots, k_m$. Let

$$\alpha = \max\{\|p_k(T)\| \mid k = 1, \ldots, N\}.$$

Given an $\epsilon > 0$ there is an m such that $\|(\alpha K)^m x_0\| < \epsilon$. Then, since $TK = KT$, we have

$$\|p_{k_m}(T)K \cdots p_{k_2}(T)K p_{k_1}(T)K x_0\| = \|p_{k_m}(T) \cdots p_{k_2}(T)p_{k_1}(T)K^m x_0\|$$
$$= \|\alpha^{-1} p_{k_m}(T) \cdots \alpha^{-1} p_{k_2}(T)\alpha^{-1} p_{k_1}(T)(\alpha K)^m x_0\|.$$

By the construction of α, $\|\alpha^{-1} p_k(T)\| \leq 1$ for all k. Therefore,

$$\|p_{k_m}(T)K \cdots p_{k_2}(T)K p_{k_1}(T)K x_0\| \leq \|(\alpha K)^m x_0\| < \epsilon.$$

This shows that given an $\epsilon > 0$ there is an element in the closed ball B of norm less than ϵ. This contradicts that $0 \notin B$. $\qquad\square$

In a sense that we have hinted at, but will not discuss, compact operators and quasinilpotent operators are related. After considering compact operators, it is therefore sensible to consider quasinilpotent operators. Read has recently constructed a quasinilpotent operator with no nontrivial subspace [103].

There are other constructed counterexamples, and other positive results. This is a very active area that we are able only to touch on here.

FIGURE 5.1. Per Enflo (r.) receiving a goose from Stanisław Mazur in 1972. The prize was offered by Mazur in 1936 for solving a problem.

FIGURE 5.2. A more recent picture of Per Enflo.

Per Enflo (Figures 5.1 and 5.2) was born on May 20, 1944, in Stockholm, Sweden.[1] His father was a surveyor, his mother an actress. Per Enflo is one of five children born to his parents. His family has been, and is, very active in music and other performing arts, and this involvement has been a strong influence in his life.

During his school years, the family moved to various places in Sweden, but Enflo enjoyed a stable, happy home life and good schooling. Around the age of eight, he became interested in both mathematics and music. These are the two subjects that he was prodigious in and to which he remains most devoted. You are reading about him because of his mathematics,

but in fact, Enflo is almost equally a musician and a mathematician.

In music, Enflo has studied piano, composition, and conducting. His first recital was given at age eleven. In 1956 and 1961 he was the winner of the Swedish competitions for young pianists. We shall not say much about his music but do mention a few recent activities. He competed in the first annual Van Cliburn Foundation's International Piano Competition for Outstanding Amateurs in 1999. During the spring of 2000, he played over half a dozen recitals.

Though devoted to both mathematics and music, it is the former that has determined where Enflo has lived. All of his academic degrees have been awarded by the University of Stockholm. Since completing his education in 1970, Enflo

---

[1] This biographical information was supplied by Per Enflo, via personal correspondence.

has held positions at the University of Stockholm, the University of California at Berkeley, Stanford University, the École Polytechnique in Paris, the Mittag-Leffler Institute and Royal Institute of Technology in Stockholm, and at the Ohio State University. Since 1989 he has held the prestigious position of "University Professor" at Kent State University.

Per Enflo is most well known for his solutions, in the 1970s, of the "approximation problem," the "basis problem," and the "invariant subspace problem." These were three fundamental and famous problems from the early days of functional analysis. Since the 1930s, many mathematicians had tried to solve them, but they remained open for about 40 years. The solutions are negative in the sense that they are solved by counterexamples; they are positive in the sense that the new methods and concepts have had a great impact on the further development of functional analysis.

The approximation problem asks whether or not every compact operator on every Banach space is the limit of finite-rank operators. The basis problem asks whether or not an arbitrary Banach space must have a Schauder basis; a sequence $\{x_k\}_{k=1}^{\infty}$ in a Banach space X is a *Schauder basis* if to each $x \in X$ there exists a unique sequence $\{a_k\}_{k=1}^{\infty}$ of complex numbers such that

$$x = \lim_{n \to \infty} \left(\sum_{k=1}^{n} a_k x_k \right).$$

Per Enflo's solution to the approximation problem also gives a counterexample to the basis problem. This work was started in 1967 and completed in 1972 and is a long story of progress and failures and of slowly developing new insights and techniques for a final success.

Arguably, his most famous mathematical contribution thus far is his solution to the invariant subspace problem. He constructed a Banach space X and a bounded linear operator $T : X \to X$ with no nontrivial invariant subspaces. The paper containing this example was published in 1987 [40], but it had existed in manuscript form for about twelve years prior to that date. The published paper is 100 pages long, and contains very difficult mathematics. His work on the invariant subspace problem was accomplished during the years 1970–1975, so one can see that the late 1960s to mid 1970s was a period of remarkable brilliance for Enflo. Enflo's counterexample, though it gives a complete solution to the invariant subspace problem, left open many doors for future research. For example, determining classes of operators that must have invariant subspaces (in the spirit of Lomonosov's result) is an active area. Enflo continues to work in this area, and, equally, some of the mathematics developed in his solution to the invariant subspace problem have led him to progress in other areas of operator theory.

The mathematical work discussed in the last few paragraphs might seem particularly abstract, but parts of the associated work have genuine applications. For example, some of the best available software algorithms for polynomial factorizations are based on ideas found in Enflo's solution to the invariant subspace problem [41]. Also, there are indications that his Banach space work may have good applications to economics.

Enflo's other important mathematical contributions include several results on general Banach space theory, and also his work on an infinite-dimensional version of Hilbert's 5th problem.

Enflo's early career as a musician is an important background both for his originality as a mathematician and for his strong interest in interdisciplinary science. He has done work in biology, on the zebra mussel invasion and phosphorus loading of Lake Erie (work funded by the Lake Erie Protection Fund). In anthropology he has worked on human evolution and has developed a "dynamic" population genetics model that lends strong support for a multiregional theory of human evolution. He has also published work in acoustics, on problems related to noise reduction.

5.5 The Spectral Theorem for Compact Hermitian Operators

There are many "spectral theorems." There is a spectral theorem for linear opera-tors on finite-dimensional spaces. In fact, you probably know this theorem. It says, informally, that any Hermitian matrix is diagonalizable. There are spectral theo-rems for a few different types of bounded linear operators on infinite-dimensional Hilbert spaces, and there are versions of the theorem for unbounded operators. In this section we will prove, as the title of the section suggests, the spectral theorem for compact Hermitian operators on Hilbert spaces.

In a broad sense, any spectral theorem says that the operator in question can be put in a diagonal form. This special form is represented as an infinite sum of operators of a more basic type. Further, the basic operators in the sum are determined by the spectrum of the original operator. Think of how the promised diagonalization of a Hermitian matrix works: You find the eigenvalues, and then the matrix can be viewed as a linear combination of matrices where each term in this representation is an eigenvalue times a diagonal matrix with 1's and 0's judiciously placed on the diagonal. More precisely, assume that T is a square matrix that is diagonalizable and let $\lambda_1, \ldots, \lambda_k$ be the distinct eigenvalues of T. Then there are square matrices of the same size, $P_1, \ldots, P_k$, that are projections ($P_i^2 = P_i$) such that

$$T = \sum_{i=1}^{k} \lambda_i P_i, \qquad I = \sum_{i=1}^{k} P_i,$$

and $P_i P_j = 0$ for all $i \neq j$. In this, P_i is the projection onto the eigenspace $\ker(\lambda_i I - T)$.

Throughout this section T will be a bounded linear operator mapping a Hilbert space H into itself. Recall that T is a compact operator if for every bounded sequence $\{x_n\}_{n=1}^{\infty}$ in H, the sequence $\{Tx_n\}_{n=1}^{\infty}$ in H has a convergent subsequence in H. Also, recall that T is Hermitian if

$$\langle Tx, y \rangle = \langle x, Ty \rangle$$

for all $x, y \in H$.

If T is both compact and Hermitian, then Theorem 5.20 gives us a real eigenvalue λ_1 of T satisfying

$$|\lambda_1| = \|T\|.$$

Let x_1 denote a corresponding unit eigenvector. Put $H_1 = H$ and

$$H_2 = \{x \in H \mid \langle x, x_1 \rangle = 0\}.$$

Then, for each $x \in H_2$,

$$\langle Tx, x_1 \rangle = \langle x, Tx_1 \rangle = \langle x, \lambda_1 x_1 \rangle = \lambda_1 \langle x, x_1 \rangle = 0,$$

and thus H_2 is a T-invariant subspace of H. The restriction $T|_{H_2}$ of T to H_2 is a compact and Hermitian operator, and if $T|_{H_2}$ is not the zero operator, we can deduce (again from Theorem 5.20) the existence of an eigenvalue and eigenvector, λ_2 and x_2, such that $x_2 \in H_2$, $\|x_2\| = 1$, and $|\lambda_2| = \|T|_{H_2}\|$. By this last equality, it should be clear that

$$|\lambda_2| \le |\lambda_1|.$$

Continue in this way to obtain nonzero eigenvalues

$$\lambda_1, \lambda_2, \ldots, \lambda_n$$

with corresponding unit eigenvectors

$$x_1, x_2, \ldots, x_n.$$

This process produces T-invariant subspaces

$$H_1 \supseteq H_2 \supseteq \cdots \supseteq H_n,$$

where

$$H_{k+1} = \{x \in H_k \mid \langle x, x_j \rangle = 0, \, j = 1, 2, \ldots, k\}.$$

Also,

$$|\lambda_k| = \|T|_{H_k}\|,$$

which shows that

$$|\lambda_1| \ge |\lambda_2| \ge \cdots \ge |\lambda_n|.$$

If $T|_{H_{n+1}} = 0$, then this process stops, and if $x \in H$, then

$$Tx = \sum_{k=1}^{n} \lambda_k \langle x, x_k \rangle x_k.$$

To see why this is the case, set

$$y_n = x - \sum_{k=1}^{n} \langle x, x_k \rangle x_k.$$

Then for each $j = 1, 2, \ldots, n$,

$$\langle y_n, x_j \rangle = \langle x, x_j \rangle - \sum_{k=1}^{n} \langle x, x_k \rangle \langle x_k, x_j \rangle = \langle x, x_j \rangle - \langle x, x_j \rangle \langle x_j, x_j \rangle = 0.$$

Therefore, $y_n \in H_{n+1}$, so that $T y_n = 0$. Consequently,

$$Tx = T y_n + T \left(\sum_{k=1}^{n} \langle x, x_k \rangle x_k \right)$$

$$= 0 + \sum_{k=1}^{n} \langle x, x_k \rangle T x_k$$

$$= \sum_{k=1}^{n} \lambda_k \langle x, x_k \rangle x_k,$$

as desired.

If the process above does not stop, then $\lim_{k \to \infty} |\lambda_k| = 0$ and

$$Tx = \sum_{k=1}^{\infty} \lambda_k \langle x, x_k \rangle x_k$$

for each $x \in H$. We point out that the λ_k's in this sum are not necessarily distinct. Why should $\lim_{k \to \infty} |\lambda_k| = 0$? Suppose that there are infinitely many distinct λ_k's and that $|\lambda_k| \geq \delta$ for some $\delta > 0$ and all $k = 1, 2, \ldots$. Then the sequence $\{\lambda_k^{-1} x_k\}_{k=1}^{\infty}$ is bounded (by δ^{-1}), and so, since T is a compact operator, the sequence $\{T(\lambda_k^{-1} x_k)\}_{k=1}^{\infty} = \{x_k\}_{k=1}^{\infty}$ has a convergent subsequence. This sequence is thus also Cauchy, which contradicts the fact that

$$\|x_k - x_j\|^2 = \langle x_k - x_j, x_k - x_j \rangle = \langle x_k, x_k \rangle - \langle x_j, x_k \rangle - \langle x_k, x_j \rangle + \langle x_j, x_j \rangle = 2$$

for all $k \neq j$. Hence, $\lim_{k \to \infty} |\lambda_k| = 0$ whenever there are infinitely many distinct eigenvalues. In this case, and again with

$$y_n = x - \sum_{k=1}^{n} \langle x, x_k \rangle x_k,$$

we have that (Exercise 4.2.1)

$$\|y_n\|^2 = \|x\|^2 - \sum_{k=1}^{n} |\langle x, x_k \rangle|^2 \leq \|x\|^2.$$

Since $y_n \in H_{n+1}$ and $|\lambda_{n+1}| = \|T|_{H_{n+1}}\|$, it is the case that

$$\|T y_n\| \leq |\lambda_{n+1}| \cdot \|y_n\| \leq |\lambda_{n+1}| \cdot \|x\|.$$

This shows that $\lim_{n \to \infty} \|T y_n\| = 0$ and hence that

$$Tx = \sum_{k=1}^{\infty} \lambda_k \langle x, x_k \rangle x_k$$

for each $x \in H$.

So far, we have produced a sequence of nonzero eigenvalues for T. Could there be any other nonzero eigenvalues of T? Suppose that λ is a nonzero eigenvalue and that x is a corresponding unit eigenvector. Then, for each $k = 1, 2, \ldots$,

$$\lambda \langle x, x_k \rangle = \langle \lambda x, x_k \rangle = \langle Tx, x_k \rangle = \langle x, Tx_k \rangle = \langle x, \lambda_k x_k \rangle = \lambda_k \langle x, x_k \rangle.$$

Since $\lambda \neq \lambda_k$, $\langle x, x_k \rangle = 0$ for each $k = 1, 2, \ldots$. Then

$$\lambda x = Tx = \sum_{k=1}^{\infty} \lambda_k \langle x, x_k \rangle x_k = 0,$$

a contradiction. Therefore, the constructed sequence $\{\lambda_k\}_{k=1}^{\infty}$ contains all of the nonzero eigenvalues of T.

A final useful observation is that the eigenspaces $\ker(\lambda_k I - T)$, $k = 1, 2, \ldots$, are each finite-dimensional. To see this, note that even though there may be infinitely many nonzero eigenvalues, a given λ_k can appear only finitely many times in the list (this is because $\lim_{k \to \infty} |\lambda_k| = 0$). Fix an index k and let $k = k(1), k(2), \ldots, k(p)$ denote the complete set of indices for which the eigenvalue $\lambda_{k(i)}$ is equal to λ_k. Thus λ_k appears exactly p times in the list. We know that the eigenvectors $x_k, x_{k(2)}, \ldots, x_{k(p)}$ are orthonormal. If $\ker(\lambda_k I - T)$ had dimension greater than p, then there would be some unit vector $x \in \ker(\lambda_k I - T)$ such that the vectors $x, x_k, x_{k(2)}, \ldots, x_{k(p)}$ were orthonormal. If $j \neq k, k(2), \ldots, k(p)$, then $\langle x, x_j \rangle = 0$ by an argument similar to many we have already seen. Thus $\langle x, x_j \rangle = 0$ for all values of j, and hence

$$\lambda_k x = Tx = \sum_{j=1}^{\infty} \lambda_j \langle x, x_j \rangle x_j = 0,$$

which contradicts $\lambda_k \neq 0$. Therefore, $\ker(\lambda_k I - T)$ must have dimension p.

We have now established the following result.

Theorem 5.23 (The Spectral Theorem for Compact Hermitian Operators).   *Let T be a nonzero, compact, and Hermitian operator on a Hilbert space H. The procedure described in the preceding paragraphs gives a sequence of nonzero real-valued eigenvalues $\{\lambda_k\}$ and a corresponding sequence of orthonormal eigenvectors $\{x_k\}$. If the sequence of eigenvalues contains infinitely many distinct values, then it tends to zero. Each nonzero eigenvalue of T appears in the sequence $\{\lambda_k\}$, and each eigenspace $\ker(\lambda_k I - T)$ is finite-dimensional. The dimension of the eigenspace $\ker(\lambda_k I - T)$ is precisely the number of times the eigenvalue λ_k appears in the sequence $\{\lambda_k\}$. Finally, for each $x \in H$, we have*

$$Tx = \sum_{k=1}^{\infty} \lambda_k \langle x, x_k \rangle x_k.$$

Let K be any closed subspace of H. By Theorem 5.16 we know that any $x \in H$ can be written in form $x = y + z$ for some $y \in K$ and $z \in K^{\perp}$. Since this decomposition is unique, we can define an operator $P : H \to H$ by $Px = y$. This operator satisfies $\|Px\| \leq \|x\|$ and hence is bounded, is linear, and satisfies

$P^2 = P$. The operator P is called the *projection* of H onto K. (See Exercise 5.5.1.) Specifically, we let P_k denote the projection of H onto the closed subspace $\ker(\lambda_k I - T)$ of H. Then the sum

$$Tx = \sum_{k=1}^{\infty} \lambda_k \langle x, x_k \rangle x_k$$

can be rewritten as

$$Tx = \sum_{[\lambda_k]} \lambda_k P_k x,$$

where the index $[\lambda_k]$ indicates that the sum is extended over *distinct* eigenvalues. This formula now gives T decomposed as a sum of more basic operators, and these operators are determined by the spectrum (specifically, by the eigenvalues) of T. We remark that there is something to prove in rearranging the sum in the statement of the theorem so that all of the terms with the same λ_k appear consecutively (see, for example, [124], Theorem II.6.9).

We remark that the Hilbert space H in the theorem is completely arbitrary. However, unless it is separable, there is no hope that the sequence of eigenvectors will be a complete orthonormal sequence for H. We might hope for this since in the finite-dimensional setting the nicest matrices are those for which there is an orthonormal basis for H consisting of eigenvectors, and because a complete orthonormal sequence replaces the finite-dimensional notion of basis when working in the infinite-dimensional setting. If H is separable, then the sequence of orthonormal eigenvectors can always be extended to a complete orthonormal sequence for H (see, for example, [129], Corollary 8.16).

The adjective "Hermitian" in Theorem 5.23 can be replaced by the adjective "normal." We have not defined this term yet (and we will not). For our purposes, it is enough to know that any compact normal operator can be written as a linear combination of two commuting compact Hermitian operators. This fact lets one deduce the spectral theorem for compact normal operators from Theorem 5.23 without too much trouble. This work was essentially done by Hilbert in 1906. Hilbert and F. Riesz soon afterwards proved their spectral theorem for bounded Hermitian (and, more generally, normal) operators. Dropping the compactness hypothesis leads to many more difficulties to overcome. Noncompact operators may not have eigenvalues, and the spectrum may be uncountable. In this case the "sum" in the decomposition is replaced by an "integral." Hilbert and Riesz's generalization is thus significantly more sophisticated, since it requires measure theory in order to discuss the integral. In the 1920s, Marshall Stone (1903–1989; U.S.A.) and John von Neumann further generalized the spectral theorem to include *unbounded* Hermitian operators. Also, there is a version of the spectral theorem for compact, but not Hermitian, operators that was obtained by F. Riesz in 1918.

Exercises for Chapter 5

Section 5.1

5.1.1 Let H be a Hilbert space and x_0 be a fixed element of H. Prove that the map taking x to $\langle x, x_0 \rangle$ is a linear operator from H to $\mathbb{C}$.

5.1.2 Consider the operator defined by

$$Tf(s) = \frac{1}{s} \int_0^s f(t)dt.$$

This is a Fredholm operator and is often referred to as the *Cesàro operator*.

   (a) Determine the kernel $k(s, t)$ of the Cesàro operator.

   (b) Prove that the Cesàro operator is linear.

5.1.3 In this section you encountered infinite matrices, perhaps for the first time. In general, matrix multiplication is not commutative. This is the case for finite as well as for infinite matrices. But multiplication for infinite matrices gets even worse. For finite matrices, there are certain situations in which multiplication is, in fact, commutative. For example, if $AB = I$ for two square matrices A and B, then A and B commute. This shows that a finite matrix with a "left inverse" also has a "right inverse" (and they are equal). For infinite-dimensional matrices, this need not be the case. Give examples of infinite matrices A and B such that $AB = I$, yet $BA \neq I$. See Exercise 5.3.3 for more on this topic.

Section 5.2

5.2.1 Prove that $\mathcal{B}(X, Y)$ is a normed linear space. That is, prove Theorem 5.3.

5.2.2 Show that the identity operator from $(\mathcal{C}([0, 1]), \| \cdot \|_\infty)$ to $(\mathcal{C}([0, 1]), \| \cdot \|_1)$ is a *bounded* linear operator, but that the identity map from $(\mathcal{C}([0, 1]), \| \cdot \|_1)$ to $(\mathcal{C}([0, 1]), \| \cdot \|_\infty)$ is *unbounded*. This phenomenon cannot happen on a Banach space; see Exercise 6.3.2.

5.2.3 Fix a continuous function $\phi : [0, 1] \to [0, 1]$ and define the *composition operator* $C_\phi : C([0, 1]) \to C([0, 1])$ by $C_\phi f(x) = f(\phi(x))$. Prove that C_ϕ is a bounded linear operator with $\|C_\phi\| \leq 1$.

5.2.4 For $\{a_n\}_{n=1}^\infty \in \ell^\infty$ define

$$T(x) = \Sigma_{n=1}^\infty x_n a_n, \quad \text{for all } x = \{x_n\}_{n=1}^\infty \in \ell^1.$$

Show that T is a bounded linear operator from ℓ^1 to $\mathbb{C}$, and compute $\|T\|$.

5.2.5 Consider an $n \times n$ matrix $A = (a_{ij})$ that satisfies $a_{ij} = \overline{a_{ji}}$ for all pairs of indices i and j. Such a matrix is called *Hermitian*. View A as a linear operator $\mathbb{C}^n \to \mathbb{C}^n$. Assume that $\mathbb{C}^n$ is endowed with the usual norm. Prove that $\|A\| = \max_k\{|\lambda_k|\}$, where λ_k are the eigenvalues of A.

5.2.6 Consider

$$A = \begin{pmatrix} a & b \\ c & d \end{pmatrix}, \quad a, b, c, d \in \mathbb{C},$$

as a linear operator from $\mathbb{C}^2$ to itself. The aim of this exercise is to show that the operator norm of A depends on the choice of norm on $\mathbb{C}^2$.

(a) Endow $\mathbb{C}^2$ with the supremum norm, $\| \cdot \|_\infty$. Show that in this case, the norm of A is given by

$$\max\left(|a| + |b|, |c| + |d|\right).$$

(b) Endow $\mathbb{C}^2$ with the 1-norm, $\| \cdot \|_1$. Show that in this case, the norm of A is given by

$$\max\left(|a| + |c|, |b| + |d|\right).$$

(c) Endow $\mathbb{C}^2$ with the 2-norm, $\| \cdot \|_2$. Consider the matrix

$$A^* = \begin{pmatrix} \bar{a} & \bar{c} \\ \bar{b} & \bar{d} \end{pmatrix},$$

and form the matrix AA^*. Show that in this case, the norm of A is given by

$$\left[\frac{1}{2}\left(\text{tr}(AA^*) + \sqrt{(\text{tr}(AA^*))^2 - 4\det(AA^*)}\right)\right]^{\frac{1}{2}}.$$

Use this to write down a formula for the norm of A as a function of its entries a, b, c, d. Here "tr" denotes trace, and "det" denotes determinant.

(d) Finally, compute these 3 operator norms of

$$A = \begin{pmatrix} 1 & 1 \\ 0 & 1 \end{pmatrix}.$$

5.2.7 Let $C^1([0, 1])$ denote the collection of all continuous functions with continuous derivative (including one-sided derivatives at the endpoints).

(a) Show that $C^1([0, 1])$, with the supremum norm, is a subspace of $C([0, 1])$ but that it is not closed.

(b) Prove that the differential operator

$$\frac{d}{dx} : C^1([0, 1]) \to C([0, 1])$$

is linear but not bounded.

Section 5.3

5.3.1 Assume that X is a normed linear space, and $S, T \in \mathcal{B}(X)$.

  (a) Prove that

$$\|ST\| \le \|S\| \cdot \|T\|.$$

  (b) Prove that $ST \in \mathcal{B}(X)$.

5.3.2 Assume that X and Y are normed linear spaces. Let K be a compact operator in $\mathcal{B}(X, Y)$. Use Theorem 2.4 to prove that $K(B)$ has compact closure for each closed ball B in X.

5.3.3 The situation discovered in Exercise 5.1.4 is somewhat redeemable. Let X be a Banach space and $T \in \mathcal{B}(X)$. Assume that there are $U, V \in \mathcal{B}(X)$ satisfying

$$UT = I = VT.$$

Prove that $U = V$. In other words, if $T \in \mathcal{B}(X)$ is both left and right invertible, then its left and right inverses must be equal.

5.3.4 In a unital normed algebra with multiplicative identity e, prove that $\|e\| \ge 1$.

5.3.5 In a normed algebra, prove that $\|a^k\| \le \|a\|^k$ for each element a in the algebra and each positive integer k.

5.3.6 Prove that $C([0, 1])$, with the supremum norm, is a unital Banach algebra.

5.3.7 Assume that X is a Banach space. In Theorem 5.6 we saw that any $S \in \mathcal{B}(X)$ that is sufficiently close to an invertible operator T is also invertible. Given that $\|S - T\| < \|T^{-1}\|^{-1}$, prove that S^{-1} satisfies

$$\|S^{-1}\| \le \frac{\|T^{-1}\|}{1 - \|T^{-1}\| \cdot \|S - T\|}.$$

5.3.8 Assume that X is a Banach space. Assume that $S, T \in \mathcal{B}(X)$ and $\lambda \in \mathbb{C} \setminus \sigma(ST)$, $\lambda \ne 0$.

  (a) Verify that

$$(\lambda I - TS)U = I = U(\lambda I - TS),$$

  where

$$U = \frac{1}{\lambda}I + \frac{1}{\lambda}T(\lambda I - ST)^{-1}S.$$

  (b) Deduce that $\sigma(ST) \cup \{0\} = \sigma(TS) \cup \{0\}$.

  (c) Use the shifts T and S as defined in Example 2 of Section 1 to show that the equality $\sigma(ST) = \sigma(TS)$ need not hold.

5.3.9 Consider the operator on $L^2((0, 1))$ defined by

$$Tf(s) = \int_0^s (s - t)f(t)dt.$$

Prove that $I - T$ is an invertible operator.

5.3.10 Let $\ell^1(\mathbb{Z})$ denote the collection of all (doubly ended) sequences $\{a_i\}_{i=-\infty}^{\infty}$ of complex numbers such that $\sum_{i=-\infty}^{\infty} |a_i| < \infty$. This space is very much like $\ell^1 = \ell^1(\mathbb{N})$. With the norm of $a = \{a_i\}_{i=-\infty}^{\infty}$ given by

$$\|a\| = \sum_{i=-\infty}^{\infty} |a_i|,$$

$\ell^1(\mathbb{Z})$ becomes a Banach space. The point of this exercise is to see that we can define a multiplication to make $\ell^1(\mathbb{Z})$ into a Banach algebra. For $a = \{a_i\}_{i=-\infty}^{\infty}$ and $b = \{b_i\}_{i=-\infty}^{\infty}$ in $\ell^1(\mathbb{Z})$, define the product $a * b$ by

$$(a * b)_i = \sum_{j=-\infty}^{\infty} a_{i-j} b_j.$$

This multiplication is called *convolution*. Show that $\ell^1(\mathbb{Z})$ is a Banach algebra.

5.3.11 Prove that every compact operator is a bounded operator.

5.3.12 Consider the Volterra operator defined on $L^2([0, 1])$ by

$$Tf(s) = \int_0^s k(s, t) f(t) dt$$

with $|k(s, t)| \le C$ for some constant C and every s and t in $[0, 1]$.

    (a) Prove that $Tf \in L^2([0, 1])$ whenever $f \in L^2([0, 1])$.

    (b) Prove that $T \in B\left(L^2([0, 1])\right)$. That is, prove that T is bounded.

    (c) Prove that T is quasinilpotent. (As it turns out, T is quasinilpotent even if the hypothesis that $|k(s, t)| \le C$ is dropped. However, the proof is substantially harder. See pages 98–99 of [58] for a proof.)

    (d) Now let $k(s, t) = 1$ for all s and t. Show that T is compact.

5.3.13 In this exercise you are asked to fill in the details of Example 4.

    (a) Show that $M_\phi \in B\left(L^2([0, 1]), \mathbb{R}\right)$, and that $\|M_\phi\| = \|\phi\|_\infty$.

    (b) Show that if $\lambda \notin f([0, 1])$, then (as you are asked to show in the same exercise) the multiplication operator $M_{(\lambda-\phi)^{-1}}$ is a bounded operator and is the inverse of $\lambda I - M_\phi$.

    (c) We have now established that $\sigma(M_\phi) = f([0, 1])$. Sometimes this set contains eigenvalues; sometimes it does not. Give examples to show this. That is, construct functions ϕ and ψ such that $\sigma(M_\phi)$ contains at least one eigenvalue and $\sigma(M_\psi)$ contains no eigenvalues at all.

5.3.14 Let W be the weighted shift in Example 6.

    (a) Show that W has no eigenvalues.

    (b) Use induction to show that $\|W^n\| \le \frac{1}{n!}, n = 1, 2, \ldots$.

5.3.15 Let X be an infinite-dimensional Banach space. (Parts (a) and (b) are related only in that they are about compact operators.)

(a) Give an example of a compact operator T such that $T^2 = 0$.

(b) Prove that if there is a positive integer n such that $T^n = I$, then T cannot be compact.

5.3.16 Assume that X is a Banach space. Prove that the collection of all compact operators forms a closed, two-sided ideal in the Banach algebra $B(X)$. Note: You may know what an "ideal" is from abstract algebra; if you do not, it is defined in the next chapter.

Warning: We end with two exercises that should not be taken lightly! In particular, they should not be considered as part of a standard "exercise set."

5.3.17 Learn enough about complex function theory to prove the results of this section.

5.3.18 Find a good reference (like [21] or [124]) and work through the proof of Theorem 5.11, which describes the spectrum of a compact operator.

Section 5.4

5.4.1 Verify that the Hermitian matrices (as defined in Exercise 5.2.5) are exactly the Hermitian operators on the finite-dimensional Hilbert space $\mathbb{C}^n$.

5.4.2 Describe the Fredholm integral operators of the first kind that are Hermitian.

5.4.3 Let K be a subspace of a Hilbert space H.

(a) Prove that $(K^\perp)^\perp = K$ whenever K is a closed subspace.

(b) Find an example to show that the equality of (a) need not hold if K is not closed.

5.4.4 Complete the proof of Theorem 5.15 by verifying the uniqueness assertion.

5.4.5 Prove Theorem 5.17.

5.4.6 It follows from Theorem 5.18 that any eigenvalue of a Hermitian operator is real-valued. Give a more direct proof of this fact.

5.4.7 Let T be the Volterra operator on $L^2([0, 1])$ defined by

$$Tf(s) = \int_0^s f(t)dt.$$

For $0 \le \alpha \le 1$ set

$$M_\alpha = \{f \in L^2([0, 1]) \mid f(t) = 0, 0 \le t \le \alpha\}.$$

(a) Prove that each M_α is an invariant subspace for T.

(b) Describe the orthogonal complement of M_α in $L^2([0, 1])$.

Section 5.5

5.5.1 Verify each of the assertions made in the paragraph that follows Theorem 5.23. That is, prove that P is bounded, linear, and satisfies $P^2 = P$.

5.5.2 In each of the following examples, explain why T is compact and Hermitian. Then give the spectral decomposition guaranteed by Theorem 5.23. That is, find the eigenvalues, the corresponding eigenspaces, and projections.

(a) The operator $T \in B(\mathbb{C}^2)$ given by the matrix

$$\begin{pmatrix} 3 & -4 \\ -4 & -3 \end{pmatrix}.$$

(b) The operator $T \in B(\ell^2)$ given by the formula

$$T(\{x_k\}_{k=1}^{\infty}) = \left\{ \frac{x_k}{k} \right\}_{k=1}^{\infty}.$$

6
Further Topics

In this chapter we present a smorgasbord of treats. The sections of this chapter are, for the most part, independent of each other (the exception is that the third section makes use of the main theorem of the second section). The sections are not uniform in length or level of difficulty. They may be added as topics for lectures, or used as sources of student projects.

The first section gives a proof of the important Weierstrass approximation theorem and of its generalization due to Marshall Stone (1903–1989; USA). We offer a proof of the latter that is relatively recent [26] and has not appeared, to our knowledge, in any text. The second section presents a theorem of René Baire and gives an application to real analysis. This material is standard in a real analysis text; we include it because we like it and because we use Baire's theorem in the third section, where we prove three fundamental results of functional analysis. In the fourth section we prove the existence of a set of real numbers that is not Lebesgue measurable. The fifth section investigates contraction mappings and how these special maps can be used to solve problems in differential equations. In the penultimate section we study the algebraic structure of the space of continuous functions. In the last section we give a very brief introduction to how some of the ideas in the text are used in quantum mechanics. Some of these topics certainly belong in a functional analysis text; some probably do not. Enjoy!

6.1   The Classical Weierstrass Approximation Theorem and the Generalized Stone–Weierstrass Theorem

The classical Weierstrass approximation theorem was first proved by Weierstrass in 1885. Several different proofs have appeared since Weierstrass's (see [54], page 266). We present the proof of it due to the analyst and probabalist Sergei Bernstein (1880–1968; Ukraine) [18].[1] His proof gives a clever, and perhaps surprising, application of probability. Marshall Stone's generalization of Weierstrass's famous theorem appeared a quarter of a century later, in [118].

Recall that $C([a, b]; \mathbb{R})$, or $C([a, b])$, denotes the Banach space of all continuous real-valued functions with norm $\|f\|_\infty = \sup\{|f(x)| \, | a \le x \le b\}$.

Theorem 6.1 (The Weierstrass Approximation Theorem).   *The polynomials are dense in $C([a, b])$. That is, given any $f \in C([a, b])$ and an $\epsilon > 0$ there exists a polynomial $p \in C([a, b])$ such that $\|f - p\|_\infty < \epsilon$.*

PROOF.   First we will show that the result is true for $a = 0$ and $b = 1$. Thus, we consider $f \in C([0, 1])$ and proceed to describe a polynomial that is close to f (with respect to the supremum norm). The polynomial we will use is a so-called *Bernstein polynomial*. The nth-degree Bernstein polynomial associated to f is defined by

$$p_n(x) = \sum_{k=0}^{n} \binom{n}{k} f\left(\frac{k}{n}\right) x^k (1-x)^{n-k},$$

where

$$\binom{n}{k} = \frac{n!}{k!(n-k)!}$$

is the binomial coefficient.

The idea for this definition comes from probability. Imagine a coin with probability x of getting heads. In n tosses, the probability of getting exactly k heads is thus

$$\binom{n}{k} x^k (1-x)^{n-k}.$$

If $f\left(\frac{k}{n}\right)$ dollars are paid when exactly k heads are thrown in n tosses, then the average dollar amount (after throwing n tosses very many times) paid when n tosses are made is

$$\sum_{k=0}^{n} \binom{n}{k} f\left(\frac{k}{n}\right) x^k (1-x)^{n-k}.$$

This expression is what we called $p_n(x)$.

---

[1]As an incidental historical remark, Bernstein's Ph.D. dissertation contained the first solution to Hilbert's 19th problem on elliptic differential equations.

We now show that given an $\epsilon > 0$ there exists n large enough so that $\|p_n - f\|_\infty < \epsilon$.

This should seem plausible: If n is very large, then we expect $\frac{k}{n}$ to be very close to x. We thus expect the average dollar amount paid, $p_n(x)$, to be very close to $f(x)$.

To prove that $\|p_n - f\|_\infty < \epsilon$ for sufficiently large n, we recall the binomial theorem:

$$(x + y)^n = \sum_{k=0}^{n} \binom{n}{k} x^k y^{n-k}. \tag{6.1}$$

If we differentiate this with respect to x and multiply both sides by x, we get

$$nx(x + y)^{n-1} = \sum_{k=0}^{n} \binom{n}{k} kx^k y^{n-k}. \tag{6.2}$$

If instead, we differentiate twice and multiply both sides by x^2, we get

$$n(n - 1)x^2(x + y)^{n-2} = \sum_{k=0}^{n} \binom{n}{k} k(k - 1)x^k y^{n-k}. \tag{6.3}$$

Equations (6.1)–(6.3), with $y = 1 - x$, read

$$1 = \sum_{k=0}^{n} \binom{n}{k} x^k (1 - x)^{n-k}, \tag{6.4}$$

$$nx = \sum_{k=0}^{n} \binom{n}{k} kx^k (1 - x)^{n-k}, \tag{6.5}$$

and

$$n(n - 1)x^2 = \sum_{k=0}^{n} \binom{n}{k} k(k - 1)x^k (1 - x)^{n-k}. \tag{6.6}$$

Therefore,

$$
\begin{aligned}
\sum_{k=0}^{n}(k - nx)^2 \binom{n}{k} x^k (1 - x)^{n-k} &= \sum_{k=0}^{n} k^2 \binom{n}{k} x^k (1 - x)^{n-k} \\
&\quad - 2 \sum_{k=0}^{n} nkx \binom{n}{k} x^k (1 - x)^{n-k} \\
&\quad + \sum_{k=0}^{n} n^2 x^2 \binom{n}{k} x^k (1 - x)^{n-k} \\
&= [nx + n(n - 1)x^2] - 2nx \cdot nx + n^2 x^2 \\
&= nx(1 - x). \tag{6.7}
\end{aligned}
$$

At this stage, you may not see where the proof is headed. Fear not, and forge ahead!

For our given f, choose $M > 0$ such that $|f(x)| \leq M$ for all $x \in [0, 1]$. Since f is continuous on $[a, b]$, it is uniformly continuous, and therefore there exists a

$\delta > 0$ such that $|f(x) - f(y)| < \epsilon$ whenever $|x - y| < \delta$ (this ϵ is the ϵ fixed at the beginning). Then

$$
\begin{aligned}
|f(x) - p_n(x)| &= \left| f(x) - \sum_{k=0}^{n} f\left(\frac{k}{n}\right) \binom{n}{k} x^k (1-x)^{n-k} \right| \\
&= \left| \sum_{k=0}^{n} \left(f(x) - f\left(\frac{k}{n}\right) \right) \binom{n}{k} x^k (1-x)^{n-k} \right| \\
&= \left| \sum_{|k-nx|<\delta n} \left(f(x) - f\left(\frac{k}{n}\right) \right) \binom{n}{k} x^k (1-x)^{n-k} \right. \\
&\quad \left. + \sum_{|k-nx|\geq \delta n} \left(f(x) - f\left(\frac{k}{n}\right) \right) \binom{n}{k} x^k (1-x)^{n-k} \right| \\
&\leq \left| \sum_{|k-nx|<\delta n} \left(f(x) - f\left(\frac{k}{n}\right) \right) \binom{n}{k} x^k (1-x)^{n-k} \right| \\
&\quad + \left| \sum_{|k-nx|\geq \delta n} \left(f(x) - f\left(\frac{k}{n}\right) \right) \binom{n}{k} x^k (1-x)^{n-k} \right|.
\end{aligned}
$$

If $|k - nx| < \delta n$, then $|x - \frac{k}{n}| < \delta$, so that $|f(x) - f\left(\frac{k}{n}\right)| < \epsilon$. Then

$$
\begin{aligned}
\left| \sum_{|k-nx|<\delta n} \left(f(x) - f\left(\frac{k}{n}\right) \right) \binom{n}{k} x^k (1-x)^{n-k} \right| & \\
\leq \sum_{|k-nx|<\delta n} \left| f(x) - f\left(\frac{k}{n}\right) \right| \binom{n}{k} x^k (1-x)^{n-k} & \\
< \epsilon \cdot \left(\sum_{|k-nx|<\delta n} \binom{n}{k} x^k (1-x)^{n-k} \right) & \\
\leq \epsilon \cdot \left(\sum_{k=0}^{n} \binom{n}{k} x^k (1-x)^{n-k} \right) = \epsilon. &
\end{aligned}
$$

If $|k - nx| \geq \delta n$, then

$$
\begin{aligned}
\left| \sum_{|k-nx|\geq \delta n} \left(f(x) - f\left(\frac{k}{n}\right) \right) \binom{n}{k} x^k (1-x)^{n-k} \right| & \\
\leq \sum_{|k-nx|\geq \delta n} \left| f(x) - f\left(\frac{k}{n}\right) \right| \binom{n}{k} x^k (1-x)^{n-k} & \\
\leq \sum_{|k-nx|\geq \delta n} \left(|f(x)| + \left| f\left(\frac{k}{n}\right) \right| \right) \binom{n}{k} x^k (1-x)^{n-k} & \\
\leq 2M \cdot \left(\sum_{|k-nx|\geq \delta n} \binom{n}{k} x^k (1-x)^{n-k} \right) &
\end{aligned}
$$

$$\leq \frac{2M}{n^2\delta^2} \cdot \left(\sum_{k=0}^{n} (k-nx)^2 \binom{n}{k} x^k (1-x)^{n-k}\right),$$

since $\dfrac{k-nx}{n\delta} \geq 1$ for these terms. By (7) the last expression is equal to

$$\frac{2M}{n^2\delta^2} nx(1-x).$$

And since $x(1-x) \leq \frac{1}{4}$ for each value of $x \in [0, 1]$, this is less than or equal to

$$\frac{M}{2n\delta^2}.$$

We have now shown that $|f(x) - p_n(x)|$ is is less than or equal to

$$\left| \sum_{|k-nx|<\delta n} \left(f(x) - f\left(\frac{k}{n}\right)\right) \binom{n}{k} x^k (1-x)^{n-k} \right|$$

$$+ \left| \sum_{|k-nx|\geq \delta n} \left(f(x) - f\left(\frac{k}{n}\right)\right) \binom{n}{k} x^k (1-x)^{n-k} \right|,$$

and that this, in turn, is less than

$$\epsilon + \frac{M}{2n\delta^2}.$$

If n is now chosen larger than $\dfrac{M}{2\delta^2\epsilon}$, we have

$$\|f - p_n\|_\infty < \epsilon + \frac{M}{2\frac{M}{2\delta^2\epsilon}\delta^2} = 2\epsilon.$$

We have now proved the Weierstrass approximation theorem for the interval $[0, 1]$ and are ready to extend this argument to an arbitrary interval $[a, b]$. The method employed here to generalize from $[0, 1]$ to $[a, b]$ is useful, and should be kept in mind. We consider any $f \in C([a, b])$, let $\epsilon > 0$ be arbitrary, and note that a and b are any two real numbers satisfying $a < b$. Define

$$g(x) = f\,(x(b-a)+a), \qquad x \in [0, 1].$$

Note that $g \in C([0, 1])$, $g(0) = f(a)$, and $g(1) = f(b)$. By the preceding argument there exists a (Bernstein) polynomial $p \in C([0, 1])$ such that $\|p-g\|_\infty < \epsilon$. Define

$$q(x) = p\left(\frac{x-a}{b-a}\right), \qquad x \in [a, b].$$

Note that q is a polynomial in $C([a, b])$, $q(a) = p(0)$, and $q(b) = p(1)$. Then

$$|f(x) - q(x)| = \left| g\left(\frac{x-a}{b-a}\right) - p\left(\frac{x-a}{b-a}\right) \right|,$$

and thus

$$\|f - q\|_\infty = \sup_{x\in[a,b]} \left\{ |f(x) - q(x)| \right\}$$

$$= \sup_{x \in [a,b]} \left\{ \left| g\left(\frac{x-a}{b-a} \right) - p\left(\frac{x-a}{b-a} \right) \right| \right\}$$

$$= \sup_{x \in [0,1]} \left\{ |g(x) - p(x)| \right\} = \|g - p\|_\infty < \epsilon.$$

This completes the proof.   □

Marshall Stone recognized that the interval, $[a, b]$ in the real line that Weierstrass used could be replaced by a more general subset of a more general metric space. In fact, he realized that the set $[a, b]$ could be replaced by any compact subset of any Hausdorff topological space. Since this book does not assume knowledge of topological spaces, we will give a proof for a compact subset of a metric space (a metric space is an example of a Hausdorff topological space). The proof goes over, verbatim, if the set X in the theorem is considered as a subset of a Hausdorff topological space.

In the theorem, we write $C(X; \mathbb{R})$ to emphasize that the functions are *real*valued. After the proof we will make a comment about the theorem for complex-valued functions.

Most proofs of the Stone–Weierstrass theorem make use of the Weierstrass approximation theorem; one attractive feature of the proof presented here is that the Weierstrass approximation theorem is not used to deduce the more general result. Hence, the Weierstrass approximation theorem is subsumed by the Stone–Weierstrass theorem. The proof we give is due to Brosowski and Deutsch [26]; we follow their presentation closely. Following the proof, we will say more about standard proofs, and also about further generalizations.

The trouble, in a general metric space, is that "polynomials" might not make sense. We observe that polynomials are exactly the functions that can be obtained from the two functions 1 and x by multiplication by a scalar, by addition, and/or by multiplication. This characterization of the polynomials is captured by the three hypotheses of the theorem.

Theorem 6.2 (The Stone–Weierstrass Theorem). *Let X be a compact metric space, and assume that $A \subseteq C(X; \mathbb{R})$ satisfies the following conditions:*

(a) *A is an algebra: If $f, g \in A$ and $\alpha \in \mathbb{R}$, then $f + g$, $f \cdot g$, and αf are all in A.*
(b) *The constant function $x \to 1$ is in A (and hence A contains all constant functions).*
(c) *A separates points: For $x \neq y \in X$ there exists an $f \in A$ such that $f(x) \neq f(y)$.*

Then A is dense in $C(X; \mathbb{R})$.

PROOF.   Brosowski and Deutsch break their proof into three parts. They prove two preliminary lemmas, and then prove the Stone–Weierstrass theorem.

The first lemma states:

Consider any point $x_0 \in X$ and any open set U_0 in X containing x_0. Then there exists an open set $V_0 \subseteq U_0$ containing x_0 such that for each $\epsilon > 0$ there exists $g \in A$ satisfying

   (i) $0 \le g(x) \le 1, x \in X$;
  (ii) $g(x) < \epsilon, \; x \in V_0$;
 (iii) $g(x) > 1 - \epsilon, \; x \in X \setminus V_0$.

To prove this lemma we first make use of hypothesis (c) to deduce the existence of a function $g_x \in A$ with $g_x(x_0) \ne g_x(x)$ for each $x_0 \in X \setminus U_0$. The function $h_x = g_x - g_x(x_0)$ is in A and satisfies $0 = h_x(x_0) \ne h_x(x)$. The function

$$p_x = \left(\frac{1}{\|h_x\|_\infty^2} \right) h_x^2$$

is also in A and satisfies $p_x(x_0) = 0$, $p_x(x) > 0$, and $0 \le p_x \le 1$.

Let $U_x = \{y \in X : p_x(y) > 0\}$. Then U_x is an open set and contains x. Since $A \setminus U_0$ is compact (by Exercise 2.1.13), it contains a finite collection of points $x_1, x_2, \ldots, x_m$ such that

$$X \setminus U_0 \subseteq \bigcup_{i=1}^{m} U_{x_i}.$$

The function

$$p = \left(\frac{1}{m} \right) \sum_{i=1}^{m} p_{x_i}$$

is in A and satisfies $0 \le p \le 1$, $p(x_0) = 0$, and $p > 0$ on $X \setminus U_0$. Since $X \setminus U_0$ is compact, there exists $0 < \delta < 1$ such that $p \ge \delta$ on $X \setminus U_0$ (see Exercise 2.1.14). The set

$$V_0 = \left\{ x \in X \,\middle|\, p(x) < \frac{\delta}{2} \right\}$$

is an open set. Further, V_0 contains the point x_0 and is contained in the set U_0.

Let k be the smallest integer greater than $\frac{1}{\delta}$. Then $k - 1 \le \frac{1}{\delta}$, and so $k < \frac{2}{\delta}$. Therefore, $1 < k\delta < 2$. Define functions

$$q_n(x) = \left(1 - p^n(x) \right)^{k^n}$$

for $n = 1, 2, \ldots$. Then $q_n \in A$, $0 \le q_n \le 1$, and $q_n(x_0) = 1$.
   If $x \in V_0$, then

$$kp(x) \le k \cdot \frac{\delta}{2} < 1.$$

The inequality proved in Exercise 6.1.2 implies that

$$q_n(x) \ge 1 - \left(kp(x) \right)^n \ge 1 - \left(k \cdot \frac{\delta}{2} \right)^n.$$

The last expression goes to 1 as $n \to \infty$.

If $x \in X \setminus U_0$, then

$$kp(x) \geq k\delta > 1,$$

and the same inequality from the exercise implies that

$$
\begin{aligned}
q_n(x) &= \frac{1}{k^n p^n(x)} \left(1 - p^n(x)\right)^{k^n} k^n p^n(x) \\
&\leq \frac{1}{k^n p^n(x)} \left(1 - p^n(x)\right)^{k^n} \left(1 + k^n p^n(x)\right) \\
&\leq \frac{1}{k^n p^n(x)} \left(1 - p^n(x)\right)^{k^n} \left(1 + p^n(x)\right)^{k^n} \\
&= \frac{1}{k^n p^n(x)} \left(1 - p^{2n}(x)\right)^{k^n} \\
&\leq \frac{1}{(k\delta)^n}.
\end{aligned}
$$

The last expression goes to 0 as $n \to \infty$.

We can therefore choose n large enough so that q_n satisfies $0 \leq q_n \leq 1$, $q_n(x) < \epsilon$ for each $x \in X \setminus U_0$, and $q_n(x) > 1 - \epsilon$ for each $x \in V_0$. Define $g = 1 - q_n$. It is left as an exercise (Exercise 6.1.3(a)) to show that g satisfies (i)–(iii).

The second lemma states:
Consider disjoint closed subsets Y and Z of X. For each $0 < \epsilon < 1$ there exists $g \in A$ satisfying

(i) $0 \leq g(x) \leq 1, x \in X$;
(ii) $g(x) < \epsilon, x \in Y$;
(iii) $g(x) > 1 - \epsilon, x \in Z$.

To prove this lemma we begin by considering the open set $U = X \setminus Z$ in X. If $x \in Y$, then $x \in U$, and the first lemma gives an open set V_x of X containing x with certain properties. Since X is compact, X contains a finite collection of points $x_1, x_2, \ldots, x_m$ such that

$$X \subseteq \bigcup_{i=1}^{m} V_{x_i}.$$

Let g_i be the function associated to V_{x_i} as given in the first lemma satisfying $0 \leq g_i \leq 1$, $g_i(x) < \frac{\epsilon}{m}$ for each $x \in V_{x_i}$, and $g_i(x) > 1 - \frac{\epsilon}{m}$ for each $x \in X \setminus U = Z$. Define $g = g_1 g_2 \cdots g_m$. It is left as an exercise (Exercise 6.1.3(b)) to show that this function satisfies (i)–(iii).

We now return to the proof of the Stone–Weierstrass theorem. Consider $f \in C(X; \mathbb{R})$. We aim to show that corresponding to any given $\epsilon > 0$ there exists an element $g \in A$ satisfying $\|f - g\|_\infty < \epsilon$. In fact, we will show that there exists an element $g \in A$ satisfying $\|f - g\|_\infty < 2\epsilon$.

Replacing f by $f + \|f\|_\infty$, we can assume that $f \geq 0$. We also assume that $\epsilon < \frac{1}{3}$. Start by choosing an integer n such that $(n - 1)\epsilon \geq \|f\|_\infty$ and define sets

$X_i, Y_i, i = 0, 1, \ldots, n$, by

$$X_i = \left\{ x \in X \mid f(x) \leq \left(i - \frac{1}{3} \right) \epsilon \right\}$$

and

$$Y_i = \left\{ x \in X \mid f(x) \geq \left(i + \frac{1}{3} \right) \epsilon \right\}.$$

Then we see that

$$X_i \cap Y_i = \emptyset,$$
$$\emptyset \subseteq X_0 \subseteq X_1 \cdots \subseteq X_n = X,$$

and

$$Y_0 \supseteq Y_1 \supseteq \cdots \supseteq Y_n = \emptyset.$$

From the second lemma we have, corresponding to each $i = 0, 1, \ldots, n$, a function $g_i \in A$ satisfying $0 \leq g_i \leq 1$, $g_i(x) < \frac{\epsilon}{n}$ for each $x \in X_i$, and $g_i(x) > 1 - \frac{\epsilon}{n}$ for each $x \in Y_i$.

Define

$$g = \epsilon \sum_{i=0}^{n} g_i.$$

This function is in A. Consider an arbitrary element $x \in X$. From the chain of subsets $\emptyset \subseteq X_0 \subseteq X_1 \cdots \subseteq X_n = X$, we see that there is an $i \geq 1$ such that $x \in X_i \setminus X_{i-1}$. For this value of i,

$$\left(i - \frac{4}{3} \right) \epsilon < f(x) < \left(i - \frac{1}{3} \right) \epsilon$$

and

$$g_j(x) < \frac{\epsilon}{n} \text{ for every } j \geq i.$$

Note also that $x \in Y_j$ for every value of $j \leq i - 2$ and thus

$$g_j(x) > 1 - \frac{\epsilon}{n} \text{ for every } j \leq i - 2.$$

These last two inequalities yield

$$g(x) = \epsilon \sum_{j=0}^{i-1} g_j(x) + \epsilon \sum_{j=i}^{n} g_j(x)$$
$$\leq i\epsilon + \epsilon(n - i + 1)\frac{\epsilon}{n}$$
$$\leq i\epsilon + \epsilon^2$$
$$< \left(i + \frac{1}{3} \right) \epsilon,$$

and, for $i \geq 2$,

$$g(x) \geq \epsilon \sum_{j=0}^{i-2} g_j(x)$$

$$\geq (i-1)\epsilon\left(1 - \frac{\epsilon}{n}\right) = (i-1)\epsilon - \frac{i-1}{n}\epsilon^2 > (i-1)\epsilon - \epsilon^2$$

$$> \left(i - \frac{4}{3}\right)\epsilon.$$

This last equality, $g(x) > \left(i - \frac{4}{3}\right)\epsilon$, is proved for $i \geq 2$; for $i = 1$ it is straightforward. Therefore,

$$|f(x) - g(x)| \leq \left(i + \frac{1}{3}\right)\epsilon - \left(i - \frac{4}{3}\right)\epsilon < 2\epsilon,$$

and the Stone–Weierstrass theorem is proved.   □

The conditions (a), (b), and (c) are usually rather easy to check if you are given a subset A of $C(X; \mathbb{R})$. Also, we point out that if $\mathbb{R}$ is replaced by $\mathbb{C}$ and the following condition (d) is added to the hypotheses, then the conclusion of the theorem still holds:

(d) If $f \in A$ then $\overline{f} \in A$.

Our proof of the Stone–Weierstrass theorem is not found in other texts. In many texts ([30] or [112], for example), a proof is given that was published in 1961 by Errett Bishop (1928–1983; USA) [19]. Bishop deduces the conclusion of the Stone–Weierstrass theorem from a powerful result now called "Bishop's theorem."

As mentioned in the paragraphs preceding Brosowski and Deutsch's proof, X can be replaced by a Hausdorff space. In what follows, we will use this more general language; if it makes you more comfortable, you may continue to think of X as a metric space. Consider a compact Hausdorff space X and a closed unital subalgebra A of $C(X; \mathbb{C})$. A subset S of X is said to be A-*symmetric* if every $h \in A$ that is real-valued on S is actually constant on S. Assume that $f \in C(X; \mathbb{C})$ and to each A-symmetric subset S of X there is a function $g_S \in A$ such that $g_S(x) = f(x)$ for all $x \in S$. Bishop's theorem asserts that with these hypotheses, f must be in A. Why is the Stone–Weierstrass theorem a consequence of Bishop's theorem? With notation as in the the statement of the Stone–Weierstrass theorem, consider $\overline{A}$, a closed unital subalgebra of $C(X; \mathbb{C})$. Let $f \in C(X; \mathbb{C})$ be arbitrary, and let S be an $\overline{A}$-symmetric subset of X. By Exercise 6.1.7, $S = \{s\}$ for some $s \in X$. Since $\overline{A}$ contains all constant functions, $\overline{A}$ contains the constant function g defined by $g(x) = f(s)$ for all $x \in X$. Clearly, then, $g(x) = f(x)$ for all $x \in S$. Bishop's theorem now implies that $f \in \overline{A}$. Since f was arbitrary, we conclude that $\overline{A}$ must contain all of $C(X; \mathbb{C})$, which is the conclusion of the Stone–Weierstrass theorem.

The next natural question is, How can one prove Bishop's theorem? The standard proof of this theorem uses ideas of Louis de Branges [24]. This approach requires sophisticated machinery (including the Hahn–Banach theorem, which will be proved later in this chapter) that we are not in a position to develop here. In 1977, Silvio Machado (1932–1981) offered a more elementary proof of Bishop's Theorem [85] (and hence of the Stone–Weierstrass theorem). In 1984, Thomas Ransford (born 1958; England) incorporated Brosowski and Deutsch's ideas and gave a shortened, simplified version of Machado's proof [102]. The only ingredient in Ransford's proof that can be thought of as "nonelementary" is Zorn's lemma. We have not yet encountered Zorn's lemma but we point out that the Hahn–Banach theorem, and hence de Branges's approach, also uses this lemma. Zorn's lemma will be discussed in Section 3 of this chapter. Reading and presenting Ransford's paper would be a nice student project.

We end with a few closing remarks. Subalgebras of $C(X; \mathbb{C})$ that separate the points of X and contain the constant function 1 (and hence all constant functions) are called *uniform algebras*. These algebras need not be closed under complex conjugation. The theory of uniform algebras is described in [48].

The classical Weierstrass approximation theorem asserts that on certain subsets of the real line, the polynomials are dense in the continuous functions. Stone's generalization aims to capture the essence of the collection of polynomials (in order to replace $\mathbb{R}$ with some more general space on which polynomials may not make sense). Another interesting way to generalize the classical theorem is as follows. Replace subsets of $\mathbb{R}$ with subsets of $\mathbb{C}$. Then polynomials make sense, and one can ask to characterize the subsets of $\mathbb{C}$ on which the polynomials are dense in the continuous functions. There are theorems, such as a famous one due to Carle Runge (1856–1927; Germany), that address this question. Runge's theorem is really a theorem of complex analysis; see page 198 of [30] for a proof. In the present context, it is interesting that the standard proof (as in [30]) can be replaced by a proof using functional analysis. This functional analytic proof is elegant, but (again!) requires the rather powerful Hahn–Banach theorem (see Section 3). For more on the connection between functional analysis and Runge's theorem, see Chapters 13 and 20 of [111]. The proof using the Hahn–Banach theorem uses the observation that the collection of (complex) differentiable functions on a compact subset of $\mathbb{C}$ forms a Banach space. There is a third proof, which uses the observation that this Banach space is, in fact, a Banach algebra. This is a proof that is more elementary in the sense that it does not require the Hahn–Banach theorem, and was first given in a nice article [50] by Sandy Grabiner (born 1939; USA).

Marshall Harvey Stone was born on April 8, 1903, in New York City (Figure 6.1). His father, Harlan Fiske Stone, was a distinguished lawyer who served on the U.S. Supreme Court, including service for five years as chief justice.

Stone attended public schools in New Jersey and graduated from Harvard University in 1922. His Ph.D., also done at Harvard, was awarded in 1926. He spent the majority of the first part of his professional life at Harvard, and then most of his career at the University of Chicago. He left Harvard to become the head of the mathematics department at Chicago. As head, he was largely responsible for turning Chicago's mathematics department into what many consider the strongest

FIGURE 6.1. Marshall Stone in 1982.

mathematics department in the United States at that time. During World War II, he was involved in secret work for the United States government.

Stone's Ph.D. thesis, *Ordinary linear homogeneous differential equations of order n and the related expansion problems*, was written under the direction of George David Birkhoff (1884–1944; U.S.A.). Over the next few years he continued this work, by studying the eigenfunctions of differential operators. This then led him to work with Hermitian operators on Hilbert spaces. Much of his work in this area was motivated by the quantum-mechanical theories developing at the same time. In particular, Stone was interested in extending Hilbert's spectral theory from bounded to unbounded operators. His work, at many times,

paralleled the work of von Neumann. In 1932, he published his classic book *Linear Transformations in Hilbert Space and their Applications to Analysis*, and he credits von Neumann and Riesz as the two primary sources of ideas for his work. It should be noted that this is the same publication year as Banach's treatise [11]; the two books are quite different in their aims and styles. Some other particularly important contributions of Stone's include his celebrated extension of Weierstrass's theorem on polynomial approximation (as discussed in Section 6.1) and his work on rings of continuous functions. The latter can be viewed as early work on commutative Banach algebras. Stone's work is characterized by brilliant combined use of ideas from analysis, algebra, and topology.

Stone had exceptional talent as a writer, which is demonstrated by his writings on many different topics. For example, his paper "The generalized Weierstrass approximation theorem" [119] and his book mentioned in the preceding paragraph are very enjoyable to read. He was interested in many things, especially in education and travel, and he wrote about these things as well as about mathematics. He authored, for example, a paper on mid twentieth century mathematics in China [120], uniting two of his interests. He traveled to many different countries, and was even shipwrecked in Antarctica. He died on January 9, 1989, shortly after becoming ill, in Madras, India.

6.2   The Baire Category Theorem with an Application to Real Analysis

After giving necessary definitions, we prove an important theorem of René Baire. We proceed to deduce from it that it is impossible for a function to be continuous at each rational point and discontinuous at each irrational point of the interval $(0, 1)$. Finally, we give another argument that no such function can exist, an argument due to Volterra. Volterra gave his proof about twenty years before the first appearance of Baire's theorem.

Throughout this section $M = (M, d)$ will denote a metric space.

A subset X in a metric space M is *nowhere dense* if $M \setminus \overline{X}$ is dense in M. Any subset of M that is a countable union of nowhere dense subsets of M is said to be of *first category* (in M). A subset of M that is not of first category is said to be of *second category*.

We use the following lemma to prove the Baire category theorem. In fact, the lemma and Baire's theorem are equivalent statements.

Lemma 6.3. *If $\{U_n\}_{n=1}^{\infty}$ is a sequence of open dense subsets of a complete metric space M, then $\bigcap_{n=1}^{\infty} U_n$ is dense in M.*

PROOF.   Consider $x \in M$ and $\epsilon > 0$. We aim to show that there exists an element $y \in \bigcap_{n=1}^{\infty} U_n$ such that $y \in B_\epsilon(x)$.

Since $\overline{U_1} = M$, there exists $y_1 \in U_1 \cap B_\epsilon(x)$. Since $U_1 \cap B_\epsilon(x)$ is open, there exists an open ball

$$B_{\epsilon_1}(y_1) \subseteq U_1 \cap B_\epsilon(x).$$

Let $\delta_1 = \min\{\frac{\epsilon_1}{2}, 1\}$.

Since $\overline{U_2} = M$, there exists $y_2 \in U_2 \cap B_{\delta_1}(y_1)$. Since $U_2 \cap B_{\delta_1}(y_1)$ is open, there exists an open ball

$$B_{\epsilon_2}(y_2) \subseteq U_2 \cap B_{\delta_1}(y_1).$$

Let $\delta_2 = \min\{\frac{\epsilon_2}{2}, \frac{1}{2}\}$.

Since $\overline{U_3} = M$, there exists $y_3 \in U_3 \cap B_{\delta_2}(y_2)$. Since $U_3 \cap B_{\delta_2}(y_2)$ is open, there exists an open ball

$$B_{\epsilon_3}(y_3) \subseteq U_3 \cap B_{\delta_2}(y_2).$$

Let $\delta_3 = \min\{\frac{\epsilon_3}{2}, \frac{1}{3}\}$.

Continuing in this way, we create a sequence $\{y_n\}_{n=1}^{\infty}$ and open balls $\{B_{\epsilon_n}(y_n)\}_{n=1}^{\infty}$ and $\{B_{\delta_n}(y_n)\}_{n=1}^{\infty}$ with $\delta_n = \min\{\frac{\epsilon_n}{2}, \frac{1}{n}\}$ satisfying

$$B_{\epsilon_n}(y_n) \subseteq U_n \cap B_{\delta_{n-1}}(y_{n-1})$$

for $n = 1, 2, \dots$. If $m > n$, then this implies that $y_m \in B_{\delta_{n-1}}(y_n)$ and thus

$$d(y_m, y_n) < \delta_n \leq \frac{1}{n}.$$

Hence $\{y_n\}_{n=1}^{\infty}$ is a Cauchy sequence. Since M is a complete metric space, there exists a $y \in M$ such that $d(y_n, y) \to 0$ as $n \to \infty$. From $y_m \in B_{\delta_n}(y_n)$ for all

$m > n$, it follows that $y \in \overline{B_{\delta_n}(y_n)}$ for all $n = 1, 2, \ldots$. Therefore,

$$y \in \overline{B_{\delta_n}(y_n)} \subseteq \overline{B_{\frac{\epsilon_n}{2}}(y_n)} \subseteq \overline{B_{\epsilon_n}(y_n)} \subseteq U_n$$

for all $n = 1, 2, \ldots$. From this string of inclusions we can conclude both

$$y \in \bigcap_{n=1}^{\infty} U_n$$

and $y \in B_{\epsilon_1}(y_1) \subseteq B_\epsilon(x)$, as desired. $\qquad\square$

Theorem 6.4 (The Baire Category Theorem). *Any nonempty, complete metric space is of second category.*

PROOF. Let M be a nonempty complete metric space that is of first category. Then M can be written as a countable union of nowhere dense subsets, $M = \bigcup_{n=1}^{\infty} A_n$. Then $M = \bigcup_{n=1}^{\infty} \overline{A_n}$ also, and from De Morgan's law we have $\emptyset = \bigcap_{n=1}^{\infty} (M \setminus \overline{A_n})$. But each set $M \setminus \overline{A_n}$ is open and dense (the latter from the definition of nowhere dense), and the lemma thus gives that $\bigcap_{n=1}^{\infty}(M \setminus \overline{A_n})$ is dense in M. This is impossible, and thus M must be of second category. $\qquad\square$

There are many applications of Baire's theorem in analysis. Our main use of it will appear in the next section, where we use it to establish fundamental results about linear operators between Banach spaces. We now give one application of it to the theory of real functions.

The function defined on $(0, 1)$ by

$$g(x) = \begin{cases} \dfrac{1}{q} & \text{if } x = \dfrac{p}{q} \text{ in reduced form,} \\ 0 & \text{if } x \text{ is irrational,} \end{cases}$$

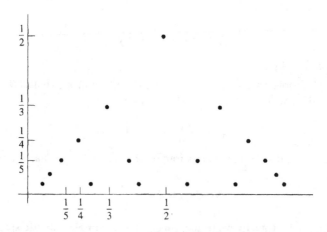

FIGURE 6.2. The graph of $g(x)$.

is a standard example of a function that is continuous at each irrational point of $(0, 1)$ and discontinuous at each rational point of $(0, 1)$ (Exercise 6.2.5). See Figure 6.2. It is natural to wonder whether there is a function defined on $(0, 1)$ that is continuous at each rational point of $(0, 1)$ and discontinuous at each irrational point of $(0, 1)$. It may seem somewhat surprising that no such function can exist. The proof of this usually given makes use of Baire's theorem, and we now supply this proof.

We start by defining, for any bounded real-valued function f defined on an open interval I, the *oscillation of f on I* by

$$\omega_f(I) = \sup\{f(x) \,|\, x \in I\} - \inf\{f(x) \,|\, x \in I\}.$$

For $a \in I$ define the *oscillation of f at a* by

$$\omega_f(a) = \inf\{\omega_f(J) \,|\, J \subseteq I \text{ is an open interval containing } a\}.$$

The connection between continuity and oscillation is made precise in the following straightforward lemma.

Lemma 6.5. *A bounded real-valued function f defined on an open interval I is continuous at $a \in I$ if and only if $\omega_f(a) = 0$ and the set $\{x \in I \,|\, \omega_f(x) < \epsilon\}$ is an open set for each $\epsilon > 0$.*

PROOF. If f is continuous at a, then given $\epsilon > 0$, there exists $\delta > 0$ such that $|f(x) - f(a)| < \epsilon$ whenever $|x - a| < \delta$. Therefore,

$$\omega_f(a) \le \omega_f\Big((a - \delta, a + \delta)\Big) < \epsilon.$$

Since this holds for every $\epsilon > 0$, we have $\omega_f(a) = 0$.

Conversely, if $\omega_f(a) = 0$, then for any given $\epsilon > 0$ there exists an open interval $J \subseteq I$ containing a such that $\omega_f(J) < \epsilon$. Since J is open, there exists a $\delta > 0$ such that the interval $(a - \delta, a + \delta)$ is contained in J. Hence $|x - a| < \delta$ implies that

$$|f(x) - f(a)| \le \omega_f\Big((a - \delta, a + \delta)\Big) \le \omega_f(J) < \epsilon,$$

as desired.

To prove the second assertion of the lemma consider $\epsilon > 0$ and $x_0 \in \{x \in I : \omega_f(x) < \epsilon\}$. Let J be an open interval containing x_0 and satisfying $\omega_f(J) < \epsilon$. For any $y \in J$, $\omega_f(y) \le \omega_f(J) < \epsilon$, and thus $J \subseteq \{x \in I : \omega_f(x) < \epsilon\}$, proving that $\{x \in I : \omega_f(x) < \epsilon\}$ is open. □

Theorem 6.6. *There is no function defined on $(0, 1)$ that is continuous at each rational point of $(0, 1)$ and discontinuous at each irrational point of $(0, 1)$.*

PROOF. Suppose, to the contrary, that such an f exists. By the lemma, the set

$$U_n = \left\{ x \in (0, 1) : \omega_f(x) < \frac{1}{n} \right\}$$

is open for each $n = 1, 2, \ldots$. By the first part of the lemma, the set $\bigcap_{n=1}^\infty U_n$ is equal to $\mathbb{Q} \cap (0, 1)$. Since the rational numbers are dense in $(0, 1)$, each U_n is also dense in $(0, 1)$. Set $V_n = (0, 1) \setminus U_n$, $n = 1, 2, \ldots$. Then $(0, 1) \setminus \mathbb{Q} = \bigcup_{n=1}^{\infty} V_n$, and each V_n is nowhere dense in $(0, 1)$. If $\{r_1, r_2, \ldots\}$ is an enumeration of the rational numbers in $(0, 1)$, then

$$(0, 1) = V_1 \cup \{r_1\} \cup V_2 \cup \{r_2\} \cup \cdots,$$

and so

$$[0, 1] = \{0\} \cup \{1\} \cup V_1 \cup \{r_1\} \cup V_2 \cup \{r_2\} \cup \cdots.$$

Each set in this union is nowhere dense, and thus $[0, 1]$ is of first category, contradicting the Baire category theorem and completing our proof.   □

As already mentioned, the nonexistence of a function continuous exactly on the rational numbers is usually deduced as a corollary to Baire's theorem. Baire's theorem appeared in 1899 [9]. Two decades before the appearance of this paper the Italian mathematician Vito Volterra gave a proof of the nonexistence of such a function; he gave this proof while he was still a student. Volterra also argues by contradiction, but he avoids altogether the somewhat sophisticated ideas of category. We first encountered Volterra's proof in William Dunham's wonderful article [37].

Following Volterra's proof, we will assume that such an f exists, and let g be defined by

$$g(x) = \begin{cases} \dfrac{1}{q} & \text{if } x = \dfrac{p}{q} \text{ in reduced form,} \\ 0 & \text{if } x \text{ is irrational.} \end{cases}$$

Consider any rational point x_0 in the open interval $(0, 1)$. By continuity of f at x_0 there exists $\delta > 0$ such that $(x_0 - \delta, x_0 + \delta) \subseteq (0, 1)$ and $|f(x) - f(x_0)| < \frac{1}{2}$ whenever $|x - x_0| < \delta$. Choose a_1 and b_1 such that $[a_1, b_1] \subseteq (x_0 - \delta, x_0 + \delta)$. Then

$$|f(x) - f(y)| \leq |f(x) - f(x_0)| + |f(x_0) - f(y)| < \frac{1}{2} + \frac{1}{2} = 1$$

for all $x, y \in [a_1, b_1]$. Next, we choose an irrational point in the open interval (a_1, b_1). By the same argument, there exist points a_1' and b_1' such that $[a_1', b_1'] \subseteq (a_1, b_1)$ and $|g(x) - g(y)| < 1$ for all $x, y \in [a_1', b_1']$. Thus, for all $x, y \in [a_1', b_1']$, we have both $|f(x) - f(y)| < 1$ and $|g(x) - g(y)| < 1$.

Repeat this argument starting with the open interval (a_1', b_1') in place of $(0, 1)$ to construct a closed interval $[a_2', b_2'] \subseteq (a_1', b_1')$ such that for all $x, y \in [a_2', b_2']$ we have both $|f(x) - f(y)| < \frac{1}{2}$ and $|g(x) - g(y)| < \frac{1}{2}$.

Keep repeating this argument to construct intervals

$$(0, 1) \supseteq [a_1', b_1'] \supseteq [a_2', b_2'] \supseteq \cdots \supseteq [a_n', b_n'] \supseteq \cdots$$

and such that for all $x, y \in [a'_n, b'_n]$ we have both $|f(x) - f(y)| < \frac{1}{2^n}$ and $|g(x) - g(y)| < \frac{1}{2^n}$. By the nested interval theorem[2] there exists exactly one point contained in all intervals $[a'_n, b'_n]$. It follows that both f and g are continuous at this point, and hence that this point is simultaneously rational and irrational. Since this is impossible, we are done.

6.3   Three Classical Theorems from Functional Analysis

In this section we present the open mapping theorem, the Banach–Steinhaus theorem, and the Hahn–Banach theorem. We also discuss their history and applications. The first was proved by Banach, the second jointly by Banach and Steinhaus, and the third was proved, independently, by Hans Hahn (1879–1934; German) and Banach. These three results are fundamental theorems of functional analysis, and it may be argued that any book purporting to be a functional analysis text must include them.

The first two can be viewed as consequences of the Baire category theorem. These two theorems have to do with linear operators between normed linear spaces. The Hahn–Banach theorem is about "linear functionals": linear mappings from a linear space into the underlying field. It is different in flavor from the other two theorems of the section, but is put in this section because these three theorems are often thought of as the "bread and butter" theorems of elementary functional analysis.

Suppose that X is a linear space, α is a scalar, and $A \subseteq X$. We will use the notation αA to denote the set $\{\alpha x \mid x \in A\}$. In particular, we note the equality of open sets: $\alpha B_\beta(x) = B_{\alpha\beta}(x)$.

Theorem 6.7 (The Open Mapping Theorem).   *Consider Banach spaces X and Y and an element $T \in \mathcal{B}(X, Y)$. If T is onto, then $T(U)$ is open in Y whenever U is open in X.*

PROOF.   We split the proof into three steps:

  (i) There exists $\epsilon > 0$ such that $B_\epsilon(0) \subseteq \overline{T(B_{\frac{1}{2}}(0))}$.
  (ii) For the $\epsilon > 0$ found in (i), we have $B_\epsilon(0) \subseteq T(B_1(0))$.
  (iii) $T(U)$ is open in Y whenever U is open in X.

To prove (i), we begin by observing that since T is onto,

$$Y = T(X) = T\left(\bigcup_{n=1}^\infty B_{\frac{n}{2}}(0)\right) = \bigcup_{n=1}^\infty T\left(B_{\frac{n}{2}}(0)\right).$$

The Baire category theorem implies that there is an integer N such that one of the sets $T\left(B_{\frac{N}{2}}(0)\right)$ is not nowhere dense in Y. By Exercise 6.2.1 the interior of

---

[2]This is a standard theorem from a first real analysis course and is an immediate consequence of the completeness of the real numbers.

$\overline{T\left(B_{\frac{N}{2}}(0)\right)}$ is not empty. We can thus find, by the definition of interior, a $y_0 \in Y$ and an $r > 0$ such that $B_r(y_0) \subseteq \overline{T\left(B_{\frac{N}{2}}(0)\right)}$.

Let $\epsilon = \frac{r}{2N}$. Note that

$$\frac{y_0}{N} + y \in \overline{T\left(B_{\frac{1}{2}}(0)\right)}$$

for each $y \in B_{2\epsilon}(0)$. Now consider any $y \in B_\epsilon(0)$. Since both $\frac{y_0}{N}$ and $-\frac{y_0}{N}$ are in $\overline{T\left(B_{\frac{1}{2}}(0)\right)}$, we have

$$
\begin{aligned}
y &= \frac{-y_0}{2N} + \left(\frac{y_0}{2N} + y\right) \\
&= \frac{-y_0}{2N} + \frac{1}{2}\left(\frac{y_0}{N} + 2y\right) \\
&\in \overline{T\left(B_{\frac{1}{4}}(0)\right)} + \overline{T\left(B_{\frac{1}{4}}(0)\right)} \\
&\subseteq \overline{T\left(B_{\frac{1}{2}}(0)\right)}.
\end{aligned}
$$

This proves (i).

To prove (ii) we consider any $y \in B_\epsilon(0)$. By (i) we can choose $x_1 \in B_{\frac{1}{2}}(0)$ such that Tx_1 and y are as close to each other as we please; choose x_1 to satisfy

$$\|y - Tx_1\|_Y < \frac{\epsilon}{2}.$$

That is, $y - Tx_1 \in B_{\frac{\epsilon}{2}}(0) = \frac{1}{2}B_\epsilon(0) \subseteq \frac{1}{2}\overline{T(B_{\frac{1}{2}}(0))} = \overline{T(B_{\frac{1}{4}}(0))}$. Now we can choose x_2 in $B_{\frac{1}{4}}(0)$ such that Tx_2 and $y - Tx_1$ are as close to each other as we please; choose x_2 to satisfy

$$\|(y - Tx_1) - Tx_2\|_Y < \frac{\epsilon}{4}.$$

Continue in this way, creating a sequence

$$x_n \in B_{2^{-n}}(0),$$

$$\left\|y - \sum_{k=1}^{n} Tx_k\right\|_Y < \frac{\epsilon}{2^n}.$$

It follows from Lemma 3.21 that $\sum_{k=1}^{\infty} x_k$ converges in X. Let x denote this infinite sum. Then

$$\|x\|_X \le \sum_{k=1}^{\infty} \|x_k\|_X < \sum_{k=1}^{\infty} 2^{-k} = 1,$$

showing that $x \in B_1(0)$. Finally, it follows from the continuity of T that $Tx = y$.

We now move on to the proof of the third part. Consider an element Tx in $T(U)$. Since x is in the open set U, there is a $\delta > 0$ such that $B_\delta(x) \subseteq U$. We will be done when we show that $B_{\delta\epsilon}(Tx) \subseteq T(U)$. To see this, let $y \in B_{\delta\epsilon}(Tx)$ and write y as $Tx + y_1$, for some $y_1 \in B_{\delta\epsilon}(0)$. Then $y_2 = \frac{y_1}{\delta}$ is in $B_\epsilon(0) \subseteq T(B_1(0))$. Write $y_2 = Tx_2$ for some $x_2 \in B_1(0)$. Then $y = Tx + \delta Tx_2 = T(x + \delta x_2)$, with $x + \delta x_2 \in U$. This completes the proof. $\square$

The next result, the Banach–Steinhaus theorem, is sometimes referred to as the *uniform boundedness principle*. In fact, there are different versions of this principle, as a perusal of functional analysis texts shows. These principles give conditions on a collection of operators under which each operator in the collection is bounded (in norm) by a single (finite) number.

Theorem 6.8 (The Banach–Steinhaus Theorem).   *Consider a Banach space X and a normed linear space Y. If $\mathcal{A} \subseteq \mathcal{B}(X, Y)$ is such that*

$$\sup\{\|Tx\|_Y \,|\, T \in \mathcal{A}\} < \infty$$

for each $x \in X$, then

$$\sup\{\|T\|_{\mathcal{B}(X,Y)} \,|\, T \in \mathcal{A}\} < \infty.$$

PROOF.   Define sets

$$E_n = \{x \in X \,|\, \|Tx\|_Y \leq n \text{ for all } T \text{ in } \mathcal{A}\} = \bigcap_{T \in \mathcal{A}} \{x \in X \,|\, \|Tx\|_Y \leq n\}.$$

Each of these sets E_n is closed, and their union $\bigcup_{n=1}^\infty E_n$ is all of X. Using Exercise 6.2.1 just as in the proof of the open mapping theorem, the Baire category theorem implies that $(E_N)^\circ \neq \emptyset$, for some integer N. By definition of interior, there exists a point x_0 and a positive number r such that $B_r(x_0) \subseteq E_N$. That is, $\|Tx\|_Y \leq N$ for each x in $B_r(x_0)$ and each $T \in \mathcal{A}$. We aim to show that there is a number K such that $\|Tx\|_Y \leq K$ for every element $x \in X$ of norm 1 and each $T \in \mathcal{A}$. For an arbitrary element x in X of norm 1, we consider the element $y = \frac{r}{2}x + x_0$. Then $y \in B_r(x_0)$, and so $\|Ty\|_Y \leq N$. Therefore,

$$\frac{r}{2}\|Tx\|_Y - \|Tx_0\|_Y \leq \left\|\frac{r}{2}Tx + Tx_0\right\|_Y = \|Ty\|_Y \leq N,$$

and hence

$$\|Tx\|_Y \leq \frac{2}{r}\left(N + \|Tx_0\|_Y\right).$$

Since the number $K = \frac{2}{r}(N + \|Tx_0\|_Y)$ is independent of x, we are done. $\square$

The Banach–Steinhaus theorem is very useful as a theoretical tool in functional analysis, and for more "concrete" applications in other areas. We will not discuss these applications.

We now move on to one of the single most important results in functional analysis: the Hahn–Banach theorem.

Let X be a linear space over $\mathbb{R}$ (respectively $\mathbb{C}$). A linear operator from X into $\mathbb{R}$ (respectively $\mathbb{C}$) is called a *linear functional*. Assume now that X is a normed linear space. The collection of all continuous linear functionals[3] on X is called the *dual*

---

[3]Recall that in this context the words *continuous* and *bounded* are interchangeable.

space of X; this space is very important, and we will only touch on its properties. The notation X^* is used to denote the dual space of X, so that X^* is shorthand for $\mathcal{B}(X, \mathbb{R})$ (respectively $\mathcal{B}(X, \mathbb{C})$). For example, the next project an interested student might take on would be to identify the dual spaces of his/her favorite Banach spaces (this material is standard in most first-year graduate functional analysis texts). The early work on dual spaces led to the idea of the "adjoint" of a linear operator, an idea that has proved extremely useful in the theory of operators on Hilbert space.

If M is a proper subspace of X, and λ is a linear functional on M taking values in the appropriate field $\mathbb{R}$ or $\mathbb{C}$, then a linear functional Λ on X is called an *extension* of λ if $\lambda(x) = \Lambda(x)$ for each $x \in M$. The Hahn–Banach theorem guarantees the extension of bounded linear functionals in a norm-preserving fashion, and it is this latter assertion — about norm preservation — that is the power of the theorem.

There are many versions of the Hahn–Banach theorem. The family of these existence theorems enjoy many, and varied, applications. For an account of the history and applications of these theorems, [96] is warmly recommended. As stated at the beginning of this section, the theorem is credited, independently, to Hahn and Banach. However, this is one of those situations in mathematics where there is another person who has not received proper credit. In this case, Eduard Helly (1884–1943; Austria) should perhaps be recognized as the originator of the theorem (see [65], [96]). Briefly, Helly proved a version of the Hahn–Banach theorem [62], roughly fifteen years before the publication of the proofs of Hahn and Banach. Helly then enlisted in the army, and was severely injured in World War I. Eventually, he returned to Vienna but then was forced to flee in 1938 to avoid persecution by the Nazis. These many years outside of the academic setting damaged his career in mathematics, and caused his earlier work to remain obscure. Incidentally, the Banach–Steinhaus theorem also appeared in Helly's 1912 paper.

The proof given here of the Hahn–Banach theorem uses the axiom of choice. The relationship between the Hahn–Banach theorem and the axiom of choice is discussed in [96]. The axiom of choice is, as the name suggests, an *axiom* of set theory. That is, it is a statement that cannot be deduced from the usual axioms of set theory. Its history is rich; for a discussion of the axiom of choice see [59], page 59. The axiom of choice is stated in the next section, where we use it to prove the existence of a nonmeasurable set. As you read, note that both of our applications of the axiom of choice are assertions of existence of something: an extension of a linear functional in the Hahn–Banach theorem and a nonmeasurable set in the next section.

As it turns out, the axiom of choice has several equivalent formulations, and it is one of these other forms that we will find most useful in our proof of the Hahn–Banach theorem. We will use a form known as Zorn's lemma. To state this lemma, we need some preliminary language. A *partially ordered set* is a nonempty set S together with a relation "$\preceq$" satisfying two conditions:

(i) $x \preceq x$ for each $x \in S$;
(ii) If $x \preceq y$ and $y \preceq z$, then $x \preceq z$.

If for any x and y in a partially ordered set S either $x \preceq y$ or $y \preceq x$, we say that S is a *totally ordered set*. Consider a subset T of a partially ordered set S. An element $x \in S$ is an *upper bound* for T if $y \preceq x$ for each $y \in T$. An element $x \in S$ is a *maximal* element if $x \preceq y$ implies $y \preceq x$. *Zorn's lemma* asserts that every partially ordered set S in which every totally ordered subset has an upper bound must contain a maximal element.

We are now ready to prove a version of the Hahn–Banach theorem having to do with *real-valued* linear functionals.

Theorem 6.9 (The Hahn–Banach Theorem (Real Case)). *Let X be a real normed linear space and let M be a subspace. If $\lambda \in M^*$, then there exists $\Lambda \in X^*$ such that $\Lambda = \lambda$ on M and $\|\Lambda\| = \|\lambda\|$.*

(It is important to keep norms straight. For example, the two norms appearing in this equality are different; the equality, more precisely written, reads: $\|\Lambda\|_{B(X,\mathbb{R})} = \|\lambda\|_{B(M,\mathbb{R})}$. We will indulge in the common practice of not writing these cumbersome subscripts.)

PROOF.   We begin by considering the real-valued function

$$p(x) = \|\lambda\| \cdot \|x\|$$

defined on all of X. Observe that

$$p(x + y) \leq p(x) + p(y) \qquad \text{and} \qquad p(\alpha x) \leq \alpha p(x)$$

for all $x, y \in X$ and $\alpha \geq 0$. Also observe that $\lambda(x) \leq p(x)$ for all $x \in M$.

Next, consider a fixed $z \in X \setminus M$. For all $x, y \in M$ we have that

$$\lambda(x) - \lambda(y) = \lambda(x - y)$$
$$\leq p(x - y) = p((x + z) + (-z - y)) \leq p(x + z) + p(-z - y).$$

Hence,

$$-p(-z - y) - \lambda(y) \leq p(x + z) - \lambda(x)$$

for all $x, y \in M$. Therefore, $y \in M$ implies that

$$s = \sup_{y \in M} \{ -p(-z - y) - \lambda(y) \} \leq p(x + z) - \lambda(x),$$

and hence that

$$s \leq \inf_{x \in M} \{ p(x + z) - \lambda(x) \}.$$

For the z specified above, define the subspace M_z of X to be the subspace generated by M and z:

$$M_z = \{ x + \alpha z \mid x \in M, \alpha \in \mathbb{R} \}.$$

Notice that the representation $w = x + \alpha z$ is unique for $w \in M_z$. Define $\overline{\lambda}(w) = \lambda(x) + \alpha s$ on M_z. Then $\overline{\lambda}$ is linear, and $\overline{\lambda}(x) = \lambda(x)$ for each $x \in M$. We have thus extended λ from M to a bigger subspace M_z of X. If M_z actually equals X, we are done (and without using Zorn's lemma!). Since $\frac{x}{\alpha} \in M$ for every $\alpha \neq 0$, we see that

$$-p(-z - \frac{x}{\alpha}) - \lambda(\frac{x}{\alpha}) \leq s \leq p(\frac{x}{\alpha} + z) - \lambda(\frac{x}{\alpha}), \qquad \alpha \neq 0.$$

From this we can deduce that $\bar{\lambda}(w) \leq p(w)$ for each $w \in M_z$ (using the first inequality if $\alpha < 0$ and the second if $\alpha > 0$).

If $M_z \neq X$, we still have work to do. Recall that we are attempting to extend λ from M to X. We could repeat the process described above, extending λ from M_z to a bigger subspace of X, but chances are, we will never reach all of X in this way. To get around this difficulty we will need to employ Zorn's lemma. Denote by S the set of all pairs (M', λ') where M' is a subspace of X containing M, and λ' is an extension of λ from M to M' satisfying $\lambda' \leq p$ on M'. Define a relation "$\preceq$" on S by

$$(M', \lambda') \preceq (M'', \lambda'')$$

if M' is a proper subspace of M'' and $\lambda' = \lambda''$ on M'. This defines a partial ordering on S. Let $T = \{(M_a, \lambda_a)\}_{a \in A}$ be a totally ordered subset of S. The pair $(\bigcup_{a \in A} M_a, \tilde{\lambda})$, where $\tilde{\lambda}(x) = \lambda_a(x)$ for $x \in M_a$, is an element of S, and is seen to be an upper bound for T. Since T was an arbitrary totally ordered subset of S, Zorn's lemma now implies that S has a maximal element, which we will call $(M_\infty, \lambda_\infty)$. Observe that λ_∞ is an extension of λ from M to M_∞ that satisfies $\lambda_\infty \leq p$ on M_∞. We aim to show that M_∞ is, in fact, all of X. If it is not, then we apply the process of extension used in passing from M to M_z to create the element $(M_{\infty z}, \lambda_{\infty z})$ of S. This element satisfies

$$(M_\infty, \lambda_\infty) \preceq (M_{\infty z}, \lambda_{\infty z}).$$

Since $(M_\infty, \lambda_\infty)$ is maximal in S, we must have that $M_\infty = M_{\infty z}$, contradicting the definition of "$\preceq$". If we now let $\Lambda = \lambda_\infty$, we have an extension of λ to all of X satisfying $\Lambda(x) \leq p(x)$ for all $x \in X$. Replacing x by $-x$ gives

$$|\Lambda(x)| \leq p(x) = \|\lambda\| \cdot \|x\|$$

for all $x \in X$, showing that $\|\Lambda\| \leq \|\lambda\|$. The only thing left to do is to show that this inequality is actually an equality. To do this, for each $\epsilon > 0$ choose $x \in M$ such that $\|x\| \leq 1$ and $|\lambda(x)| > \|\lambda\| - \epsilon$. Then $|\Lambda(x)| > \|\lambda\| - \epsilon$ also and, consequently, $\|\Lambda\| \geq \|\lambda\|$, completing the proof. $\square$

We now prove a complex version of the Hahn–Banach theorem. The proof of the complex case was first given in 1938 [20]. The proof follows from the real case. In fact, most of the work is already done in the real case, and it seems surprising that over a decade passed between the appearance of the proofs of the two cases, especially in light of the explosive development of functional analysis at the time. We are now considering a complex linear space X, a subspace M of X, and a complex-valued bounded linear functional λ defined on M. We wish to extend this functional to one defined on all of X in a way that does not force the norm of the functional to grow. To see how this can be done, define real-valued functions λ_1 and λ_2 by

$$\lambda(x) = \lambda_1(x) + i\lambda_2(x)$$

for $x \in M$. You should check that λ_1 and λ_2 are real linear, and note that λ_1 satisfies

$$|\lambda_1(x)| \leq |\lambda(x)| \leq \|\lambda\| \cdot \|x\|$$

for each $x \in X$. If we view X as a linear space over $\mathbb{R}$, the first version of the Hahn–Banach theorem tells us that there is a real-valued bounded linear functional Λ_1 on X satisfying $\Lambda_1(x) = \lambda_1(x)$ for each $x \in M$ and $\|\Lambda_1\| \leq \|\lambda\|$. Next, define Λ on X by

$$\Lambda(x) = \Lambda_1(x) - i\Lambda_1(ix).$$

This is our desired extension, and it will not be hard to show that it has all the desired properties. First, you should check that it is a complex linear map on X. Since for $x \in M$,

$$\lambda_1(ix) + i\lambda_2(ix) = \lambda(ix) = i\lambda(x) = -\lambda_2(x) + i\lambda_1(x)$$

and λ_1 and λ_2 are real-valued, we have that

$$-\lambda_2(x) = \lambda_1(ix).$$

Thus,

$$\begin{aligned}
\lambda(x) &= \lambda_1(x) + i\lambda_2(x) \\
&= \lambda_1(x) - i\lambda_1(ix) \\
&= \Lambda_1(x) - i\Lambda_1(ix) \\
&= \Lambda(x)
\end{aligned}$$

for each $x \in M$, showing that Λ extends λ. It should be clear that $\|\lambda\| \leq \|\Lambda\|$, and we will be done when we show that this is actually an equality. For $x \in X$ write the complex number $\Lambda(x)$ in polar form $re^{i\theta}$ with nonnegative r and real θ. Then

$$|\Lambda(x)| = e^{-i\theta}\Lambda(x) = \Lambda(e^{-i\theta}x),$$

showing that $\Lambda(e^{-i\theta}x)$ is real and hence equal to its real part, $\Lambda_1(e^{-i\theta}x)$. Therefore,

$$|\Lambda(x)| = \Lambda_1(e^{-i\theta}x) \leq \|\Lambda_1\| \cdot \|e^{-i\theta}x\| \leq \|\lambda\| \cdot \|x\|,$$

implying that $\|\Lambda\| \leq \|\lambda\|$, as desired.

The Hahn–Banach theorem has many applications, both in functional analysis and in other areas of mathematics. Its applications within functional analysis focus on separation properties, and the interested reader should follow up by reading any one of the functional analysis texts in the References. We end this section with a brief discussion of the theorem's use in another area of mathematics.

One of the most important problems in an area of mathematics called potential theory is the so-called Dirichlet problem. Potential theory is a branch of partial differential equations and has its roots in problems of the 1700s such as the problem of determining gravitational forces exerted by bodies of various shapes (it was already known that the earth is some sort of ellipsoid). See Section 22.4 of [76] for more of this history.

We consider an open, bounded, *connected* (that is, it cannot be written as the disjoint union of two open subsets) set U in $\mathbb{R}^n$. The *boundary* ∂U of U is the set $\overline{U} \cap (\overline{\mathbb{R} \setminus U})$. Given a continuous function f defined on ∂U, the Dirichlet problem is to find a continuous function u defined on all of $\overline{U}$ that is a solution to the Laplacian equation

$$\frac{\partial^2 u}{\partial x_1^2} + \cdots + \frac{\partial^2 u}{\partial x_n^2} = 0$$

(on U) and meets the additional requirement that u take on the same values that f does on ∂U. Usually, if such a u exists, it is not too hard to show that it is the unique solution to the Dirichlet problem (for a specified U and f). Showing that a solution exists at all is harder (much harder), and various techniques can be used. One way to show this existence is to use the Hahn–Banach theorem. Further details can be found on page 155 of [47].

6.4 The Existence of a Nonmeasurable Set

The primary goal of this section is to show that there is a subset of $\mathbb{R}$ that is not Lebesgue measurable. The example can easily be adapted to give a subset of $\mathbb{R}^n$ that is not Lebesgue measurable. The example given here is a modification of one given by Giuseppe Vitali (1875–1932; Italy) [126].

The facts we need about Lebesgue measure are the following:

(i) It is *translation invariant*. That is, $m(E) = m(x + E)$ for each $x \in \mathbb{R}$ and $E \in \mathcal{M}$. Here, the set $x + E$ is defined by

$$x + E = \{x + y \mid y \in E\}.$$

(ii) $m([0, 1)) = 1$.
(iii) It is, as is every measure, countably additive.

The second follows from the definition of m; the third is proved in Theorem 3.6. The first has not yet been mentioned. We point out that a set $E \subseteq \mathbb{R}$ is measurable if and only if $x + E$ is measurable, and now (i) follows from the next theorem.

Theorem 6.10. *Lebesgue outer measure m^* is translation invariant on $2^{\mathbb{R}}$.*

PROOF.   For an interval $I = (a, b), (a, b], [a, b]$, or $[a, b), m^*(I) = m(I) = b - a$ by definition, and it should be clear that $m^*(x + I) = m^*(I)$ for each $x \in \mathbb{R}$. Let $E \subseteq \mathbb{R}$ be arbitrary and cover E with a countable number of intervals I_n,

$$E \subseteq \bigcup_{n=1}^{\infty} I_n.$$

Then

$$x + E \subseteq \bigcup_{n=1}^{\infty}(x + I_n),$$

and we have

$$m^*(x + E) \le \sum_{n=1}^{\infty} m(x + I_n) = \sum_{n=1}^{\infty} m(I_n).$$

Taking the infimum over all such covers of E, we get

$$m^*(x + E) \le m^*(E).$$

Then also

$$m^*(E) \le m^*(-x + (x + E)) \le m^*(x + E). \qquad \square$$

We also will use the *axiom of choice*: Let I be any nonempty set. If $\{A_i : i \in I\}$ is a nonempty family of pairwise disjoint sets such that $A_i \ne \emptyset$ for each $i \in I$, then there exists a set $E \subseteq \bigcup_{i \in I} A_i$ such that $E \cap A_i$ consists of exactly one element for each $i \in I$. The axiom of choice is discussed in further detail in the preceding section. If you have not yet read that section, you should read the discussion on the axiom of choice (and the equivalent Zorn's lemma) found there before proceeding.

We now turn our attention to showing that there exists a set $E \subseteq \mathbb{R}$ that is not Lebesgue measurable. We start by defining an equivalence relation $\sim$ on $(0, 1)$ by $x \sim y$ if $x - y \in \mathbb{Q}$. The equivalence classes play the role of the A_i's in the axiom of choice as stated above. Therefore, we conclude the existence of a set E in $(0, 1)$ consisting of exactly one element of each equivalence class.

We will show that E cannot be Lebesgue measurable. Assume, to the contrary, that E is Lebesgue measurable. Let $r_1, r_2, \ldots$ be an enumeration of the rational numbers in $(-1, 1)$. Define sets

$$E_n = \{x + r_n \mid x \in E\}, \qquad n = 1, 2, \ldots.$$

Since m is translation invariant, $m(E_n) = m(E)$ for each n. Note that each E_n is contained in the interval $(-1, 2)$, and thus

$$\bigcup_{n=1}^{\infty} E_n \subseteq (-1, 2).$$

Also, notice that

$$(0, 1) \subseteq \bigcup_{n=1}^{\infty} E_n.$$

To see that this is the case choose any $x \in [0, 1)$ and let y be the unique element of E that is equivalent to x. Then $x - y \in \mathbb{Q} \cap (-1, 1)$ and thus must be one of the rationals r_k for some k. In this case, $x = y + r_k \in E_k \subseteq \bigcup_{n=1}^{\infty} E_n$. Finally, we note that $E_n \cap E_m = \emptyset$ for $n \ne m$. To see that this is the case, consider an element $x \in E_n \cap E_m$. Then $x = y + r_n = z + r_m$ for some $y, z \in E$. Then

$y - z = r_m - r_n \in \mathbb{Q}$, which shows that $y \sim z$ and $y \neq z$ (since $n \neq m$). This is impossible, since E contains precisely one element from each equivalence class.

Combining the observations of the previous paragraph with (i)–(iii) above, we get

$$3 = m((-1, 2)) \geq m\left(\bigcup_{n=1}^{\infty} E_n\right) = \sum_{n=1}^{\infty} m(E_n) = \sum_{n=1}^{\infty} m(E),$$

and therefore $m(E) = 0$. On the other hand,

$$1 = m((0, 1)) \leq m\left(\bigcup_{n=1}^{\infty} E_n\right) = \sum_{n=1}^{\infty} m(E_n) = \sum_{n=1}^{\infty} m(E),$$

implying that $m(E) > 0$. Clearly, we cannot have both, and we must thus conclude that E is not Lebesgue measurable.

This proof uses the axiom of choice. Interestingly, one *must* use the axiom of choice to prove the existence of sets that are not Lebesgue measurable. This follows from results of Robert Solovay [115].

6.5   Contraction Mappings

Fixed point theorems have many applications in mathematics and are also used in other areas, such as in mathematical economics (see, for example, [22], [92]). We mentioned the Schauder fixed point theorem in the proof of Theorem 5.22 about invariant subspaces. Most theorems that ensure the existence of solutions of differential, integral, and operator equations can be reduced to fixed point theorems. The theory behind these theorems belongs to topology and makes use of ideas such as continuity and compactness. Two of the most important names associated to this broad area are Henri Poincaré (1854–1912; France) and Luitzen Brouwer (1881–1966; Netherlands). In this section we will prove a fixed point theorem known as Banach's contraction mapping principle and study applications of it to differential equations. This type of application is of an aesthetically appealing nature: The result may be known and may be provable via the methods of "hard analysis," but the techniques of functional analysis reveal a beautifully enlightening proof. The proof of our theorem is a generalization of an analytic technique due to (Charles) Emile Picard (1856–1941; France).

We begin with a differential equation subject to a boundary condition:

$$\frac{dy}{dx} = \phi(x, y),$$
$$y(x_0) = y_0.$$

"Finding a solution" means to construct a (necessarily continuous) function $y(x)$ that passes through the point (x_0, y_0) and has slope $\phi(x, y)$ near x_0. This solution, if it exists, will thus be an element of $C([a, b])$ for some closed interval $[a, b]$ containing x_0.

Given the above system, how do we know whether a solution exists at all? If one exists, can we tell whether it is the only one? Consider the following "easy" example:

$$\frac{dy}{dx} = y^{\frac{2}{3}},$$
$$y(0) = 0.$$

This, as you can easily check, has solutions

$$y_1(x) = 0 \quad \text{and} \quad y_2(x) = \frac{1}{27}x^3.$$

Therefore, uniqueness does not always follow from existence. We shall later see conditions ensuring uniqueness.

In 1820, Cauchy proved the first uniqueness and existence theorems for a system of type

$$\frac{dy}{dx} = \phi(x, y),$$
$$y(x_0) = y_0.$$

However, he imposed severe restrictions on ϕ, and the proof was unnecessarily complicated. There subsequently followed improvements on Cauchy's theorem and proof, including an improvement of the proof due to Picard. The result we present next is a general fixed point theorem, the proof of which uses Picard's iterative method that he employed in his version of Cauchy's theorem. We will then rephrase the problem about differential equations into the language of the fixed point theorem. Rephrasing the differential equations problem in the language of functional analysis yields a remarkably simple result with a powerful conclusion. This rephrasing in a more general setting also greatly increases the scope of applications.

At this juncture we must remember that a solution to our system is an element of $C([a, b])$, and that $C([a, b])$ endowed with the supremum norm

$$\|f\|_\infty = \sup\{|f(x)| \mid a \le x \le b\}$$

is a complete metric space.

Let $X = (X, d)$ be a metric space. A map $T : X \to X$ is a *contraction* if there exists $M \in [0, 1)$ such that $d(Tx, Ty) \le Md(x, y)$ for all $x, y \in X$.

Here is the main result of this section:

Theorem 6.11 (The Banach Contraction Mapping Principle). *Let X be a complete metric space and T a contraction $X \to X$. Then there exists a unique point $x \in X$ with $Tx = x$.*

PROOF.   Choose any $x_0 \in X$. Put

$$x_1 = Tx_0,$$
$$x_2 = Tx_1 (= T^2 x_0),$$

$$\vdots$$

$$x_n = T^n x_0.$$

We will show that this defines a Cauchy sequence $\{x_n\}$. Then, since X is complete, we know that this sequence converges, say to x. We will finish by showing that x is a fixed point of T and that it is the only fixed point of T.

Let M be as in the definition above and note that

$$d(x_{n+1}, x_n) \le M^n d(x_1, x_0).$$

This is left as Exercise 6.5.1.

Then, if $m > n$,

$$d(x_m, x_n) \le d(x_m, x_{m-1}) + d(x_{m-1}, x_{m-2}) + \cdots + d(x_{n+1}, x_n)$$

$$\le (M^{m-1} + M^{m-2} + \cdots + M^n) d(x_1, x_0)$$

$$\le \left(\frac{M^n}{1-M}\right) d(x_1, x_0). \quad < \varepsilon \qquad [[\text{tends to } 0)]$$

$$\text{sequence is}$$

Since

$$\text{Cauchy}$$

$$\frac{M^n}{1-M} \to 0$$

as $n \to \infty$, we see that $\{x_n\}_{n=1}^{\infty}$ is Cauchy. Let $x = \lim_{n \to \infty} x_n$ and notice that

$$d(Tx, x) \le d(Tx, Tx_n) + d(Tx_n, x) \le Md(x, x_n) + d(x_{n+1}, x).$$

Since both $Tx = x$ $-$ fixed Point

$$d(x, x_n) \to 0 \quad \text{and} \quad d(x_{n+1}, x) \to 0$$

as $n \to \infty$, we see that $d(Tx, x) = 0$. In other words, $Tx = x$. To complete the proof it remains to be shown that x is the only fixed point of T. Suppose that $Ty = y$ and that $x \ne y$. Then $d(x, y) > 0$, and thus

not some $-$
$Tx = x$

$$d(x, y) = d(Tx, Ty) \le Md(x, y) < d(x, y),$$

a contradiction. Therefore, it must be the case that $x = y$. This completes the proof. $\qquad \square$

We now return to our differential equation with boundary condition:

$$\frac{dy}{dx} = \phi(x, y),$$

$$y(x_0) = y_0.$$

As you should verify, this is equivalent to the integral equation

$$y(x) = y_0 + \int_{x_0}^{x} \phi(t, y(t)) dt.$$

Define a map

$$T : C([a, b]) \to C([a, b])$$

by $Tf = g$, where

$$g(x) = y_0 + \int_{x_0}^{x} \phi(t, f(t))dt.$$

Any solution of the original differential system is a solution to the integral equation, which in turn is a fixed point of the map T. Let us now assume that ϕ satisfies a "Lipschitz condition" in the second variable.[4] That is, we assume that there exists a positive number K such that

$$|\phi(x, y) - \phi(x, z)| \le K|y - z|$$

for all $x \in [a, b]$ and all $y, z \in \mathbb{R}$. In this case,

$$
\begin{aligned}
|(Tf)(x) - (Tg)(x)| &= \left| \int_{x_0}^{x} [\phi(t, f(t)) - \phi(t, g(t))]dt \right| \\
&\le \int_{x_0}^{x} |\phi(t, f(t)) - \phi(t, g(t))|dt \\
&\le \int_{x_0}^{x} K|f(t) - g(t)|dt \\
&\le \int_{x_0}^{x} Kd(f, g)dt \\
&\le K(b - a)d(f, g).
\end{aligned}
$$

Since this holds for all $x \in [a, b]$, we have

$$d(Tf, Tg) \le K(b - a)d(f, g)$$

for all $f, g \in C([a, b])$. From this, we see that the map T is a contraction as long as $K(b - a) < 1$. We thus have the following corollary to Banach's theorem:

Theorem 6.12. *Let notation be as in the preceding discussion. If there exists a $K > 0$ with $K(b - a) < 1$, then there exists a unique $f \in C([a, b])$ such that $f(x_0) = y_0$ and $f'(x) = \phi(x, f(x))$ for all other $x \in [a, b]$.*

In some simple situations, Picard's method of successive approximation can actually be used to construct the unique solution to a differential equation with boundary condition. For example, consider

$$f'(x) = x + f(x),$$
$$f(0) = 0.$$

This system is equivalent to the integral equation

$$f(x) = \int_{0}^{x} (t + f(t))dt.$$

---

[4]Named in honor of Rudolf Otto Sigismund Lipschitz (1832–1903; Königsberg, Prussia, now Kaliningrad, Russia).

Thus we put

$$(Tf)(x) = y_0 + \int_{x_0}^{x} \phi(t, f(t))dt,$$

where $y_0 = 0$, $x_0 = 0$, and $\phi(t, f(t)) = t + f(t)$. Does ϕ satisfy a Lipschitz condition? We have

$$|\phi(x, y) - \phi(x, z)| = |(x + y) - (x + z)| = |y - z|,$$

which is less than or equal to $K|y - z|$ if we put $K = 1$. Since we need $K(b-a) < 1$ in order to apply our theorem, we just need to restrict attention to an interval $[a, b]$ of length less than one containing $x_0 = 0$. Then our theorem implies the existence and uniqueness of a solution; what is this solution? Let us choose, quite naively, $f_0(x) = 0$. Then

$$f_1(x) = (Tf_0)(x) = \int_0^x (t + f_0(t))dt = \frac{1}{2}x^2,$$

$$f_2(x) = \frac{1}{2}x^2 + \frac{1}{6}x^3,$$

$$\vdots$$

$$f_n(x) = \frac{1}{2!}x^2 + \frac{1}{3!}x^3 + \cdots + \frac{1}{(n+1)!}x^{n+1}.$$

From this we see that $f_n(x) \to f(x) = e^x - x - 1$, and sure enough, $e^x - x - 1$ solves the original system.

There are problems with using Picard's iteration scheme. First, it may be the case that the iterates cannot be solved for with elementary functions. Or even if they can theoretically be solved for, they may be too hard to calculate. Second, even if we "have" the iterates, to figure out what they converge to may be very difficult. The power of this theorem, even though it is a "constructive" proof, is for ensuring existence and uniqueness. The theorem is also useful when one is interested in using a computer to get numerical approximations to a solution. The computational side of this problem is one that we will not go into at all, but is worthy of the interested reader's further investigation (see [92]).

6.6   The Function Space $C([a, b])$ as a Ring, and its Maximal Ideals

In this short section we show that the maximal ideals of the ring $C([a, b])$ are in one-to-one correspondence with the points of $[a, b]$. We start with definitions of rings and ideals. The intent of this section is to study the *algebraic structure* of function spaces. Though this may be of more interest to students who have studied abstract algebra, the material found here is self-contained and requires no background not already found in this text.

A *ring* is a nonempty set $\mathcal{R}$ together with two binary operations, "+" and "·", such that for all a, b, c in $\mathcal{R}$:

1. $a + b$ is in $\mathcal{R}$;
2. $a + b = b + a$;
3. $(a + b) + c = a + (b + c)$;
4. There is an element 0 in $\mathcal{R}$ such that $a + 0 = 0 + a = a$;
5. There is an element $-a$ in $\mathcal{R}$ such that $a + (-a) = 0$;
6. $a \cdot b$ is in $\mathcal{R}$;
7. $(a \cdot b) \cdot c = a \cdot (b \cdot c)$;
8. $a \cdot (b + c) = a \cdot b + a \cdot c$ and $(b + c) \cdot a = b \cdot a + c \cdot a$.

If, in addition, there is an element 1 in $\mathcal{R}$ such that $a \cdot 1 = 1 \cdot a = a$ for each $a \in \mathcal{R}$, $\mathcal{R}$ is said to be a *ring with identity* (or *ring with unit*). The use of the word ring in this section is different than the usage in the context of measure theory.

We really have no need for this definition, except to notice that $C([a, b])$ is a ring with identity. You are asked to prove this in Exercise 6.6.1.

A nonempty subset $\mathcal{J}$ of a ring $\mathcal{R}$ is called an *ideal* of $\mathcal{R}$ if

1. a and b both in $\mathcal{J}$ imply $a + b$ also in $\mathcal{J}$;
2. a in $\mathcal{J}$ and r in $\mathcal{R}$ imply that both ar and ra are in $\mathcal{J}$.

An ideal $\mathcal{J}$ of a ring $\mathcal{R}$ is a *proper* ideal if $\mathcal{J} \neq \mathcal{R}$, and is a *maximal* ideal if $\mathcal{J} = \mathcal{M}$ for every proper ideal $\mathcal{M}$ satisfying $\mathcal{J} \subseteq \mathcal{M}$. Recall from the Section 5.3 what a Banach algebra is, and that $C([a, b])$ is one. An ideal in a Banach algebra is, among other things, a subspace, and it makes sense to ask whether a given ideal is closed in the algebra.

Theorem 6.13. *A maximal ideal in a Banach algebra is always a closed ideal.*

PROOF. Let $\mathcal{J}$ be a maximal ideal. We first observe that the set of invertible elements in the Banach algebra is open (see the discussion following the proof of Theorem 5.6). Since $\mathcal{J}$ must have empty intersection with the set of invertible elements (Exercise 6.6.2), $\mathcal{J}$ cannot be dense. Therefore, $\overline{\mathcal{J}}$ is not the entire algebra. Also, it is straightforward to check that $\overline{\mathcal{J}}$ is an ideal. Since $\mathcal{J}$ is maximal, it must be the case that $\mathcal{J} = \overline{\mathcal{J}}$, as desired. ☐

In general, it can be quite a hard problem to identify the closed ideals of a Banach algebra. However, it is sometimes feasible to characterize the closed ideals that are maximal ideals. Our next theorem does just that for the Banach algebra $C([a, b])$. It must be said that this example is just the tip of a huge iceberg. First, the interval $[a, b]$ can be replaced by a much more general type of topological space (a compact Hausdorff space X). The proof in this case depends on a theorem, Urysohn's lemma,[5] from topology that is outside of the scope of this text. But even this result about $C(X)$ is far from the whole story; one might start by considering, for

---

[5] Due to Pavel Urysohn (1898–1924; Ukraine).

example, other types of functions in place of the continuous functions considered here. See [124], Sections VII.3–VII.5 for much more on this.

We define, for each $x \in [a, b]$, the set

$$\mathcal{J}_x = \{f \in C([a, b]) \,\big|\, f(x) = 0\}.$$

We are now ready to prove the main theorem of this section.

Theorem 6.14. *Each ideal $\mathcal{J}_x$ is a maximal ideal of $C([a, b])$, and moreover, every maximal ideal of $C([a, b])$ is of this form. Finally, $\mathcal{J}_x = \mathcal{J}_y$ if and only if $x = y$.*

PROOF.   Exercise 6.6.4 shows that $\mathcal{J}_x$ is a proper ideal of $C([a, b])$ for each $x \in [a, b]$. Suppose that there is a proper ideal $\mathcal{J}$ satisfying $\mathcal{J}_x \subseteq \mathcal{J}$. For $x \in [a, b]$ define $\lambda_x : C([a, b]) \to \mathbb{R}$ by $\lambda_x(f) = f(x)$. Since

$$\lambda_x(f + \alpha g) = (f + \alpha g)(x) = f(x) + \alpha g(x) = \lambda_x(f) + \alpha \lambda_x(g)$$

for all $f, g \in C([a, b])$ and each real number α, it follows that λ_x is a linear functional. Since λ_x is not identically zero, $\mathcal{J}_x = \ker \lambda_x$ is a proper subspace of $C([a, b])$. Choose any element $f \in C([a, b]) \setminus \mathcal{J}_x$. For any $g \in C([a, b])$, the element

$$g - \frac{\lambda_x(g)}{\lambda_x(f)} f$$

is in $\mathcal{J}_x$. So g is in the subspace generated by $\mathcal{J}_x$ and f; since g was arbitrary, the subspace generated by $\mathcal{J}_x$ and f is all of $C([a, b])$. This shows that $\mathcal{J}_x$ is a maximal subspace and hence must be a maximal ideal (any ideal is a subspace, so if $\mathcal{J}_x$ cannot even fit inside a proper subspace, there is no hope that it might fit inside a proper ideal).

We have now shown that each ideal of the form $\mathcal{J}_x$ is a maximal ideal. We aim to show that in fact, these are the only maximal ideals. Assume that $\mathcal{J}$ is a proper ideal. We want to show that there exists $x \in [a, b]$ such that $\mathcal{J} \subseteq \mathcal{J}_x$. Suppose, to the contrary, that $\mathcal{J} \nsubseteq \mathcal{J}_x$ for every $x \in [a, b]$. Then, for every $x \in [a, b]$, there is a function $f_x \in \mathcal{J}$ with $f_x(x) \neq 0$. Each of these functions is continuous, and thus there exist open intervals U_x on which f_x^2 is a strictly positive function. Notice that these open intervals form an open cover for the compact set $[a, b]$. Thus there exists a finite number of points $x_1, \ldots, x_n$ in $[a, b]$ such that the function

$$f = f_{x_1}^2 + \cdots + f_{x_n}^2$$

is strictly positive on all of $[a, b]$. By construction, $f \in \mathcal{J}$ and f is invertible, contradicting the result of Exercise 6.6.2. Thus, $\mathcal{J} \subseteq \mathcal{J}_x$ for some $x \in [a, b]$, and if $\mathcal{J}$ is maximal, $\mathcal{J} = \mathcal{J}_x$.

The final assertion of the theorem is left as Exercise 6.6.5. $\square$

6.7   Hilbert Space Methods in Quantum Mechanics

Quantum mechanics attempts to describe and account for the properties of molecules and atoms and their constituents. The first attempts, in large part due to Niels Bohr (1885–1962; Denmark), had limited success and are now known as the "old quantum theory." The "new quantum theory" was developed around 1925 by Werner Heisenberg (1901–1976; Germany) and Erwin Schrödinger (1887–1961; Austria), and later was extended by Paul Dirac (1902–1984; England). The story of the history and development of quantum theory is extremely interesting, for many different reasons, and includes deep philosophical questioning, personal and political intrigue, as well as fascinating mathematics and physics. There are many very good books on the subject that address these different facets, and it would be difficult to give a short recommended reading list. It is the aim of this section to give some inkling of how Hilbert space and operator theory can be used in quantum mechanics. Our treatment is not rigorous. Good references that contain more details on what we introduce here include [39], [43], [113].

For simplicity, we restrict ourselves to a consideration of one-dimensional motion. That is, we consider a particle (such as an electron) moving along a straight line, so that there is a function $f(x, t)$ of position x (on that straight line) and time t such that the probability that the particle is in the interval $[a, b]$ at time t is given by

$$\int_a^b |f(x, t)|^2 dx.$$

The (complex-valued) function f is called the *state function* for the particle. It should be clear that we expect

$$\int_{-\infty}^\infty |f(x, t)|^2 dx = 1,$$

for each fixed value of t. We now consider t to be fixed, and write $f(x)$ in place of $f(x, t)$. In summary, the state of a quantum particle in one dimension is a function f in the Hilbert space $L^2(\mathbb{R})$ satisfying $\|f\|^2 = \int_{-\infty}^\infty |f(x)|^2 dx = 1$.

The particle's position, x, is an example of an *observable*, that is, a quantity that can be measured. Other observables useful in quantum mechanics include the particle's momentum and energy. We will study position and momentum, and discuss Heisenberg's uncertainty principle. Throughout this section we are taking *Planck's constant*, $\hbar$, to be 1.

The *Fourier transform*[6] of any $L^2(\mathbb{R})$ function f is another square integrable function $\hat{f}$, given by

$$\hat{f}(w) = \frac{1}{\sqrt{2\pi}} \int_{-\infty}^\infty f(t) e^{-iwt} dt.$$

---

[6]The Fourier transform is a very useful object and not beyond the level of this book. A strong temptation to include more on this topic was resisted. Further investigation of the Fourier transform's properties and applications would make a nice project.

It is not obvious that this makes sense for all $L^2(\mathbb{R})$ functions; we refer the interested reader to Chapter 7 of [43]. In addition to the fact that the Fourier transform makes sense, we will use one of its most fundamental properties, without supplying a proof, that

$$\|f\|_2 = \|\hat{f}\|_2$$

for each f. This equation is called the *Plancherel identity*, named in honor of Michel Plancherel (1885–1967; Switzerland).

If w denotes the momentum of the particle, then the Fourier transform of the state function can be used to give the probability that w is in the interval $[a, b]$. The probability that the momentum of the particle is in $[a, b]$ is given by

$$\int_a^b |\hat{f}(w)|^2 dw.$$

If $\bar{x}, \bar{w}$ denote the average, or *mean*, values of position x and momentum w, respectively, then

$$\bar{x} = \int_{-\infty}^{\infty} x|f(x)|^2 dx, \qquad \bar{w} = \int_{-\infty}^{\infty} w|\hat{f}(w)|^2 dw,$$

and the *variance* of each is given by

$$\sigma_x^2 = \int_{-\infty}^{\infty} (x - \bar{x})^2 |f(x)|^2 dx, \qquad \sigma_w^2 = \int_{-\infty}^{\infty} (w - \bar{w})^2 |\hat{f}(w)|^2 dw.$$

Figure 6.3 gives some illustration of what the size of the variance tells us.

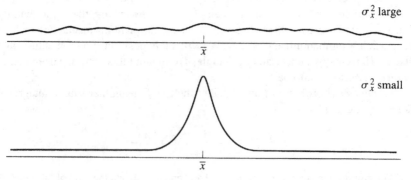

FIGURE 6.3. $|f(x)|^2$ will resemble the top graph if σ_x^2 is big, and the bottom graph if σ_x^2 is small.

Heisenberg's uncertainty principle says in this context that σ_x^2 and σ_w^2 cannot be "small" simultaneously. Specifically,

$$\sigma_x^2 \cdot \sigma_w^2 \geq \frac{1}{4}.$$

From Figure 6.3, this means, informally, that position and momentum cannot be "localized" simultaneously.

We justify this inequality for $\overline{x} = 0 = \overline{w}$ (this assumption is not very restrictive). Define operators M (for "multiplication") and D (for "differentiation") on $L^2(\mathbb{R})$ by

$$Mf(x) = xf(x) \qquad \text{and} \qquad Df(x) = f'(x).$$

We point out that these are not defined on all of $L^2(\mathbb{R})$, but for our superficial treatment we will not worry about this and will always take for granted that f is a member of the "right" domain. It should be clear that

$$\sigma_x^2 = \|Mf\|^2.$$

Also,

$$\sigma_w^2 = \int_{-\infty}^{\infty} w^2 |\hat{f}(w)|^2 dw = \int_{-\infty}^{\infty} |\widehat{Df}(w)|^2 dw = \|Df\|^2$$

(in this, the first equality should be clear; a justification of the second is asked for in Exercise 6.7.3; the third follows from the Plancherel identity). Since

$$(xf(x))' = f(x) + xf'(x),$$

we have that

$$D(Mf) = f + M(Df),$$

or, in "operator" form,

$$DM - MD = I$$

on the intersection of the subspaces of $L^2(\mathbb{R})$ on which M and D are defined. It is straightforward to see that M is Hermitian, that is, $\langle Mf, g \rangle = \langle f, Mg \rangle$, for all $f, g \in L^2(\mathbb{R})$. It is also true, but not as easy to show, that $\langle DMf, f \rangle = -\langle Mf, Df \rangle$ for all $f \in L^2(\mathbb{R})$. As you can check, this boils down to showing that

$$\int_{-\infty}^{\infty} x(|f(x)|^2)' dx = -\int_{-\infty}^{\infty} |f(x)|^2 dx,$$

and you can find a proof of this equality in Section 2.8 of [39]. Then, using the Cauchy–Schwarz inequality (Theorem 1.3),

$$\begin{aligned}
1 = \|f\|^2 &= \langle f, f \rangle \\
&= \left\langle (DM - MD)f, f \right\rangle \\
&= \langle DMf, f \rangle - \langle MDf, f \rangle \\
&= -\langle Mf, Df \rangle - \langle Df, Mf \rangle
\end{aligned}$$

$$\leq 2\big|\langle Mf, Df\rangle\big|$$
$$\leq 2\|Mf\| \cdot \|Df\|$$
$$= 2\sigma_x \cdot \sigma_w,$$

which yields the desired result that $1 \leq 4\sigma_x^2\sigma_w^2$.

We pointed out above that the operators of quantum mechanics we have discussed are not defined on all elements of the Hilbert space $L^2(\mathbb{R})$. In fact, the setting of bounded operators on Hilbert space is *not* appropriate for this context. That the operators satisfy $DM - MD = I$ is critically important, and this next result tells us that this could not happen in the $B(H)$ setting.

Theorem 6.15. *There do not exist bounded linear operators S and T on any Hilbert space that satisfy $ST - TS = I$.*

PROOF.   Suppose, to the contrary, that there is a Hilbert space H and operators $S, T \in B(H)$ that satisfy $ST - TS = I$. We will prove, using induction, that

$$nT^{n-1} = ST^n - T^n S$$

for each positive integer n. The case $n = 1$ is our hypothesis. Assume that it holds for $n > 1$. Then

$$\begin{aligned}
(n + 1)T^n &= nT^{n-1}T + T^n I \\
&= (ST^n - T^n S)T + T^n(ST - TS) \\
&= ST^{n+1} - T^n ST + T^n ST - T^{n+1} S \\
&= ST^{n+1} - T^{n+1} S.
\end{aligned}$$

Therefore,

$$nT^{n-1} = ST^n - T^n S$$

holds for each positive integer n. Recalling that the operator norm is submultiplicative, an application of the triangle inequality yields

$$n\|T^{n-1}\| \leq 2\|S\| \cdot \|T\| \cdot \|T^{n-1}\|$$

for each n. This tells us that either $\|T^{n-1}\| = 0$ for some n, or that $n \leq 2\|S\| \cdot \|T\|$ for all n. Since the latter cannot happen, we have that $\|T^{n-1}\| = 0$ for some value of n. Therefore, $T^{n-1} = 0$. Since

$$nT^{n-2} = ST^{n-1} - T^{n-1} S,$$

we deduce that $T^{n-2} = 0$. We repeat this argument n times and ultimately deduce that $I = 0$, a clear contradiction. $\square$

In finite dimensions there is an alternative, and basic, proof for this theorem: the trace of any matrix of form $ST - TS$ is zero, while the trace of the $n \times n$ identity matrix is n. There are, necessarily infinite, matrices that satisfy the equation

$ST - TS = I$; for example, take

$$S = \begin{pmatrix} 0 & 1 & 0 & 0 \dots \\ 0 & 0 & 2 & 0 \dots \\ 0 & 0 & 0 & 3 \dots \\ \vdots & \vdots & \vdots & \ddots \end{pmatrix} \quad \text{and } T = \begin{pmatrix} 0 & 0 & 0 & 0 \dots \\ 1 & 0 & 0 & 0 \dots \\ 0 & 1 & 0 & 0 \dots \\ \vdots & \vdots & \vdots & \ddots \end{pmatrix}.$$

Notice, though, that S does not define a bounded operator on the sequence space ℓ^2. See [31] for more on the equation $ST - TS = I$ and its role in the matrix mechanics of Heisenberg, Born, and Jordan, and for more on algebraic structures in quantum mechanics.

The uncertainty principle has other interpretations. For example, $f(t)$ might represent the amplitude of a signal (like a sound wave) at time t. We have changed the name of the independent variable from x to t for what we hope is an obvious reason.

EXAMPLE 1.   Fix real numbers θ and a, and let

$$f(t) = \begin{cases} \frac{1}{\sqrt{2a}} e^{i\theta t} & \text{if } t \in (-a, a), \\ 0 & \text{otherwise.} \end{cases}$$

In Exercise 6.7.2 you are asked to show that $\int_{-\infty}^{\infty} |f(t)|^2 dt = \int_{-a}^{a} |f(t)|^2 dt = 1$ and that

$$\hat{f}(w) = \frac{1}{\sqrt{\pi a}} \frac{\sin a(w - \theta)}{w - \theta}$$

(Figure 6.4).

Since $\int_{-\infty}^{\infty} |\hat{f}(w)|^2 dw$ must be 1, we observe that if a is large (that is, if the time duration of $f(t)$ is big), then the frequencies of f are near to θ. Likewise, if a is small, then the frequencies are spread out (Figure 6.5).

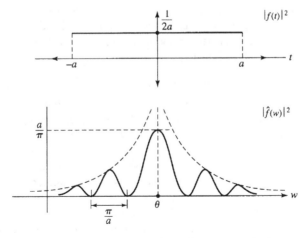

FIGURE 6.4. The area under each curve must be 1.

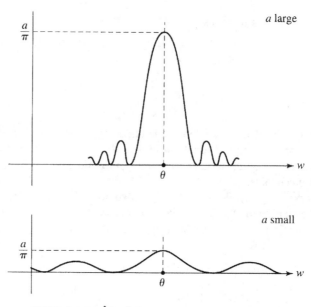

FIGURE 6.5. $|\hat{f}(w)|^2$, for large and small values of a.

We now return to the general case where $f(t)$ represents the amplitude of a signal at time t. Let a and b be positive numbers and set

$$\alpha^2 = \int_{-a}^{a} |f(t)|^2 dt$$

and

$$\beta^2 = \int_{-b}^{b} |\hat{f}(w)|^2 dw.$$

Observe that the ordered pair (α, β) is in the unit square $[0, 1] \times [0, 1]$. If $\alpha = 1$ (as is the case in Example 1), then the signal is "time-limited" (the signal is confined to the interval $[-a, a]$); $\beta = 1$ means that the signal is "band-limited." Can the ordered pair (α, β) be anywhere in the unit square? As it turns out, there is a positive number $\lambda_1 < 1$ such that

$$\arccos \alpha + \arccos \beta \geq \arccos \sqrt{\lambda_1}. \tag{6.8}$$

This assertion is an "uncertainty principle." It tells us that there are some ordered pairs (α, β) in the square that are not allowed. For example, since $\lambda_1 < 1, \alpha = \beta = 1$ cannot be achieved (for any values of a and b). This version of the uncertainty principle was proved during the years 1961–1962 by three mathematicians working at Bell Labs (see Section 2.9 of [39] for a complete reference to this theorem and for its proof). Some questions are apparent. What is this number λ_1 (and why the funny name λ_1)? As you will read, λ_1 is a function of a and b. For a given pair a and b, and hence a specified λ_1, why must α and β satisfy (6.8)? And can we identify the portion of the unit square that the points (α, β) fill in? To begin to address these questions, consider a and b as fixed positive numbers and consider the two closed subspaces M and N of $L^2(\mathbb{R})$ defined by

$$M = \{f \in L^2(\mathbb{R}) \,|\, f(t) = 0 \text{ for } t \notin [-a, a]\}$$

and

$$N = \{f \in L^2(\mathbb{R}) \,|\, \hat{f}(w) = 0 \text{ for } w \notin [-b, b]\}.$$

M is the class of time-limited functions, while N is the class of band-limited functions. Next, we consider two linear operators $T_{t\ell}$ and $T_{b\ell}$ on $L^2(\mathbb{R})$ defined by

$$T_{t\ell}f(t) = \begin{cases} f(t) & \text{if } t \in [-a, a], \\ 0 & \text{otherwise,} \end{cases}$$

and

$$T_{b\ell}f(t) = \frac{1}{\sqrt{2\pi}} \int_{-b}^{b} \hat{f}(w) e^{iwt} dw.$$

These operators are the projections from $L^2(\mathbb{R})$ onto M and N, respectively. We are interested in the operator

$$T = T_{b\ell} T_{t\ell}.$$

This operator is given by the formula (Exercise 6.7.4)

$$Tf(t) = \int_{-a}^{a} \frac{\sin b(t - s)}{\pi(t - s)} f(s) ds.$$

As it turns out, T has a countable number of real eigenvalues. The number λ_1 in (6.8) is the largest, and these eigenvalues satisfy

$$1 > \lambda_1 \geq \lambda_2 \geq \cdots \geq 0.$$

The number λ_1 is a function of the values of a and b only (because $T_{t\ell}$ and $T_{b\ell}$ are) and, in fact, depends only on the value of the product ab. As this product gets large, λ_1 approaches 1, as in Figure 6.6. As $ab \to \infty$, (6.8) thus implies that more and more points (α, β) are allowable; this is as one might expect from the definitions of α and β.

We have now described what λ_1 is. Next, let us think a bit about why (6.8) might be true. A consequence of (6.8) is that as one of α or β nears 1, the other must get smaller. Certainly, the restriction

$$\alpha^2 + \beta^2 \leq 1$$

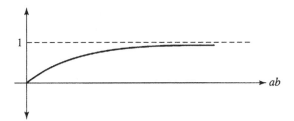

FIGURE 6.6. λ_1 as a function of ab.

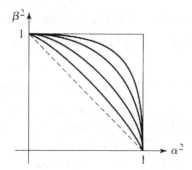

FIGURE 6.7. The curve $\arccos \alpha + \arccos \beta = \arccos \sqrt{\lambda_1}$, for various values of ab. As $ab \to 0$ this curve approaches the diagonal. Points beneath the curve satisfy (6.8).

would keep α and β from being near 1 simultaneously. In Exercise 6.7.5 you are asked to show that (6.8) is true whenever $\alpha^2 + \beta^2 \leq 1$. Are there any allowable points (α, β) inside the unit square but outside the circle $\alpha^2 + \beta^2 = 1$? That is, can one find $f \in L^2(\mathbb{R})$ that gives (α, β) outside the circle? The answer is yes if and only if (α, β) satisfies (6.8). The rest of the proof of the Bell Labs result involves showing that for each given a and b (so that λ_1 is determined), $f \in L^2(\mathbb{R})$ implies that (6.8) holds, and conversely, if α and β satisfy (6.8), then there exists f in $L^2(\mathbb{R})$ such that

$$\alpha^2 = \int_{-a}^{a} |f(t)|^2 dt \qquad \text{and} \qquad \beta^2 = \int_{-b}^{b} |\hat{f}(w)|^2 dw.$$

Finally, Figure 6.7 shows the curve $\arccos \alpha + \arccos \beta = \arccos \sqrt{\lambda_1}$, for various values of the product ab.

Of all the individuals profiled in this book, **John von Neumann** probably enjoys the greatest degree of name recognition (Figure 6.8). In fact, I am fairly confident that he is the only one for whom an obituary appeared in *Life* magazine (February 1957). Also, an interview with his (second) wife appeared in *Good Housekeeping* (September 1956).

He was born on December 28, 1903, in Budapest, Hungary. His father was a banker in a well-to-do Jewish family. Von Neumann's brilliance was apparent even at a very young age, and there are many stories about his precociousness and the exceptional mental capabilities that

remained with him throughout his life. It is said that he had a photographic memory. While this is impressive and interesting, it is not what he is remembered for. Throughout his life he could grasp very difficult concepts extremely quickly. In addition to being very sharp of mind, von Neumann worked tirelessly. He published approximately 60 articles in pure mathematics and 20 in physics. Altogether he published over 150 papers, most of the rest on applications to economics and computer science.

Trained as a chemist as well as a mathematician, von Neumann was well prepared for the scientific career that he

FIGURE 6.8. John von Neumann.

FIGURE 6.9. Stamp in honor of von Neumann.

would ultimately have. He attended good schools, and was awarded the Ph.D. from the University of Budapest in 1926. His thesis was about set theory. He worked in Germany until 1930, working mostly on the new quantum-mechanical theory and operator theory. He extended Hilbert's spectral theory from bounded to unbounded operators. This work paralleled, in large part, Stone's work at the time, but the two worked independently. Von Neumann published his great book uniting quantum mechanics and operator theory, *Mathematische Grundlagen der Quantenmechanik*, in 1932. Notice that in this same year both Banach's and Stone's books also appeared.

In 1930 von Neumann came to the United States, becoming one of the original six members of the Institute for Advanced Study at Princeton. As mentioned already, his early work focused on set theory, quantum mechanics, and operator theory. His famous proof of the "ergodic theorem" came in the early 1930s. The techniques that he developed in this context served

him later when he studied rings of operators, which became his focus later in the 1930s. "Rings of operators" are now called "operator algebras"; an important subclass of these are the "von Neumann algebras." In 1933, building on Haar's work on measures, von Neumann solved an important special case of the fifth of Hilbert's 23 problems.[7]

Around 1940, von Neumann changed the focus of his work from pure to applied mathematics. During World War II he did much work for a variety of government and civil agencies. He wrote extensively on topical subjects, including ballistics, shock waves, and aerodynamics. In addition to the over 150 published papers, he wrote many more that remain unpublished for security reasons. Von Neumann held strong political views, and was very much involved with the controversial scientific politics of World War II and the subsequent Cold War. He was one of the key players in the creation of the atomic bomb at the Los Alamos Scientific Laboratory, and later served on the Atomic Energy Commission (appointed by President Eisenhower in 1955). We will not go into his politics here; there exist extensive accounts of this in the literature. The book by Macrae [87] and its bibliography make a good starting point.

---

[7] These problems are described in the biographical material on Hilbert.

Also post-1940, he worked on mathematical economics. He is credited with the first application of game theory to economics. His theory of mathematical economics is based on his *minimax theorem*, which he had proved much earlier, in 1928 [97]. This theory was laid out in *The Theory of Games and Economic Behavior*, written jointly with Oskar Morgenstern (1902–1977; Germany) and published in 1944. Their book is now a classic.

In the years following World War II, von Neumann devoted considerable attention to the development of the modern computer. He was interested in every aspect of computing. He made significant contributions in several different areas, including parallel processing and errors involved with large computations (specifically, inverting large matrices and Monte Carlo methods). He also worked on questions about weather forecasting, and his work has had an impact on this field. In a more philosophical area, he drew an analogy between the computer and the human nervous system.

John von Neumann died on February 8, 1957, in Washington, D.C.

Exercises for Chapter 6

Section 6.1

6.1.1 Describe the Bernstein polynomials for the two functions $f(x) = x$ and $g(x) = x^2$ on $[0, 1]$.

6.1.2 Prove that $(1 + x)^n \geq 1 + nx$ for all $x \geq -1$ and each positive integer n.

6.1.3 (a) Complete the proof of the first lemma used in the proof of the Stone–Weierstrass theorem by showing that the described g satisfies (i)–(iii).

(b) Complete the proof of the second lemma used in the same theorem.

6.1.4 In the Weierstrass approximation theorem, it is crucial that the interval is compact, as addressed in Stone's generalization. One of our favorite noncompact subsets of $\mathbb{R}$ is $\mathbb{R}$ itself. Show that uniform polynomial approximation is not always guaranteed on $\mathbb{R}$. Here are two approaches: (i) Come up with a function in $C(\mathbb{R})$ that cannot be uniformly approximated by polynomials; (ii) show that the uniform limit of polynomials $\mathbb{R} \to \mathbb{R}$ is still a polynomial.

6.1.5 Deduce the Weierstrass approximation theorem from the Stone–Weierstrass theorem (thus showing that the former is indeed a special case of the latter).

6.1.6 Consider the collection $\mathcal{P}_e$ of all even polynomials.

(a) Use the Stone–Weierstrass theorem to show that $\mathcal{P}_e$ is dense in $C([0, 1])$.

(b) Explain why the Stone–Weierstrass theorem cannot be used to show that $\mathcal{P}_e$ is dense in $C([-1, 1])$.

(c) The following question remains: Is $\mathcal{P}_e$ dense in $C([-1, 1])$? Answer this question, and prove your assertion.

6.1.7 Let A be as in the hypotheses of Bishop's theorem.

(a) Prove that functions of the form $f + \overline{f}$, for $f \in A$, are real-valued and separate the points of X.

(b) Deduce that an $\overline{A}$-symmetric subset of X must be a singleton.

6.1.8 Find (in another book) another proof of the Weierstrass approximation theorem. Write up your own account of this proof. For example, if you have read the final section of the chapter, you may want to find the proof that makes use of the Fourier transform.

Section 6.2

6.2.1 Show that X is nowhere dense in M if and only if its closure has empty interior, $(\overline{X})^\circ = \emptyset$.

6.2.2 Show that any finite subset of $\mathbb{R}$ is nowhere dense in $\mathbb{R}$. Give examples to show that a countable set can be nowhere dense in $\mathbb{R}$, but that this is not always the case.

6.2.3 Let $f(x) = \sin\left(\frac{1}{x}\right)$ for $x \neq 0$, and $f(0) = 0$. Compute $\omega_f(0)$. How does this tie in with what you know about the continuity of this function?

6.2.4 First category sets are in some sense "small," while second category sets are "large." Another notion of set size that we have discussed is "measure zero": "Small" sets are of measure zero, while "large" sets are not. Is there a connection between these two ways of describing the size of a set? The answer is no. Please show this by doing two things:

(a) Describe a set that is of first category, but not of measure zero.

(b) Describe a set that is of measure zero, but not of first category.

6.2.5 Prove that the function defined on $(0, 1)$ by

$$g(x) = \begin{cases} \frac{1}{q} & \text{if } x = \frac{p}{q} \text{ in reduced form,} \\ 0 & \text{if } x \text{ is irrational,} \end{cases}$$

is continuous at each irrational point of $(0, 1)$ and discontinuous at each rational point of $(0, 1)$.

6.2.6 The goal of this exercise is to show that "most" continuous functions are nowhere differentiable. Let E_n denote the set of all $f \in C([0, 1])$ for which there exists an $x_f \in [0, 1]$ satisfying

$$|f(x) - f(x_f)| \leq n|x - x_f|$$

for every $x \in [0, 1]$.

(a) Show that E_n is nowhere dense in $C([0, 1])$. To do this, approximate $f \in C([0, 1])$ by a piecewise linear function g whose pieces each have slope $\pm 2n$. Then, if $\|h - g\|_\infty$ is sufficiently small, the function h cannot be in E_n.

(b) Deduce from (a) that the nowhere differentiable functions are of second category in $C([0, 1])$.

Section 6.3

6.3.1 Let X and Y be Banach spaces. By considering the right shift on ℓ^2 into itself defined by $S(x_1, x_2, \ldots) = (0, x_1, x_2, \ldots)$, we see that the property of being one-to-one is not enough to guarantee that a bounded linear operator be invertible (compare what happens in the finite-dimensional case). However, if a one-to-one bounded linear operator is also onto, then it must be invertible. Use the open mapping theorem to prove that the inverse is also bounded.

6.3.2 Let X be a Banach space in the two different norms $\| \cdot \|_1$ and $\| \cdot \|_2$.

(a) Show that if there exists a constant M satisfying $\|x\|_1 \leq M\|x\|_2$ for all $x \in X$, then the two norms are equivalent.

(b) Does the result of (a) contradict the result of Exercise 5.2.2? Explain your answer to this question.

Section 6.4

6.4.1 Show that $x \sim y$ if $x - y \in \mathbb{Q}$ defines an equivalence relation on the open interval $(0, 1)$.

Section 6.5

6.5.1 Give an inductive proof of the inequality $d(x_{n+1}, x_n) \leq M^n d(x_1, x_0)$, $n = 1, 2, \ldots$, and thereby complete the proof of the contraction mapping theorem.

6.5.2 Show that the method of successive approximations applied to the differential equation $f' = f$ with $f(0) = 1$ yields the usual formula for e^x.

6.5.3 For each of the following sets give an example of a continuous mapping of the set into itself that has no fixed point:

(a) $\mathbb{R}$.

(b) $(0, 1]$.

(c) $[-10, -7] \cup [2, 4]$.

6.5.4 Give an example of a mapping of $[0, 1]$ into itself that is not continuous and has no fixed points.

6.5.5 Consider the function $f : \mathbb{R} \to \mathbb{R}$ given by

$$f(x) = \begin{cases} x + e^{\frac{-x}{2}} & \text{if } x \geq 0, \\ e^{\frac{x}{2}} & \text{if } x < 0. \end{cases}$$

Show that $|f(x) - f(y)| < |x - y|$ for all x and y, yet f has no fixed point. Does this contradict the contraction mapping principle? Explain.

Section 6.6

6.6.1 For f and g in $C([a, b])$ define $f + g$ and $f \cdot g$ by

$$(f + g)(x) = f(x) + g(x) \qquad \text{and} \qquad (f \cdot g)(x) = f(x) \cdot g(x).$$

Show that $C([a, b])$ is a ring with identity.

6.6.2 Prove that a proper ideal of a ring with identity contains no invertible elements.

6.6.3 If $\mathcal{J}$ is an ideal in a Banach algebra, then $\overline{\mathcal{J}}$ is also an ideal. (In this statement, "Banach algebra" can be replaced by "normed ring.")

6.6.4 Prove that $\mathcal{J}_x$ is a proper ideal of $C([a, b])$.

6.6.5 Prove that for $x, y \in [a, b]$, $\mathcal{J}_x = \mathcal{J}_y$ if and only if $x = y$. (In proving this, you may find yourself assuming that $x \neq y$ and constructing an actual function f that is in one of the ideals but is not in the either. It is this step that requires Urysohn's lemma when $[a, b]$ is replaced by a compact Hausdorff space; see the paragraph preceding Theorem 6.14. In fact, in that case f is not actually constructed but only its existence implied.)

Section 6.7

6.7.1 Fix a positive real number a and let $\chi_{[-a,a]}$ denote the characteristic function

$$\chi_{[-a,a]}(t) = \begin{cases} 1 & \text{if } t \in [-a, a], \\ 0 & \text{otherwise.} \end{cases}$$

Show that the Fourier transform of $\chi_{[-a,a]}$ is given by

$$\widehat{\chi}_{[-a,a]}(w) = \frac{2 \sin(aw)}{w}.$$

6.7.2 As in Example 1, fix real numbers θ and a, and let

$$f(t) = \begin{cases} \frac{1}{\sqrt{2a}} e^{i\theta t} & \text{if } t \in (-a, a), \\ 0 & \text{otherwise.} \end{cases}$$

Show that $\int_{-\infty}^{\infty} |f(t)|^2 dt = 1$ and

$$\hat{f}(w) = \frac{1}{\sqrt{\pi a}} \frac{\sin a(w - \theta)}{w - \theta}.$$

Use that $e^{i\theta} = \cos \theta + i \sin \theta$ for every real number θ.

6.7.3 In the section we claimed that

$$\sigma_w^2 = \int_{-\infty}^{\infty} w^2 |\hat{f}(w)|^2 dw = \int_{-\infty}^{\infty} |\widehat{Df}(w)|^2 dw = \|Df\|^2.$$

The point of this exercise is to verify the middle equality, and hence all three equalities. Note that

$$\widehat{Df}(w) = \frac{1}{\sqrt{2\pi}} \int_{-\infty}^{\infty} f'(t) e^{-iwt} dt.$$

Use integration by parts to show that this equals $i w \hat{f}(w)$, and hence that the equation holds.

6.7.4 Verify the formula

$$Tf(t) = \int_{-a}^{a} \frac{\sin b(t-s)}{\pi(t-s)} f(s) ds$$

for $T = T_{b\ell} T_{t\ell}$, as used in the section.

6.7.5 Show that (6.8) (in the text of the section) is true whenever $\alpha^2 + \beta^2 \leq 1$.

Appendix A
Complex Numbers

In the main body of the text we consider linear spaces over two fields: $\mathbb{R}$ and $\mathbb{C}$. The reader must, therefore, be familiar with the arithmetic properties of the complex numbers. These properties are discussed in this appendix. The study of *functions* of a complex variable is rich, and is the subject for a different course. There are many excellent introductory texts about functions of a single complex variable.

We assume that the reader is familiar with properties of the real numbers, specifically, that the reader knows what is meant when one says that $\mathbb{R}$ is an "ordered field" and that a "completeness property" is satisfied in $\mathbb{R}$.

The set of real numbers is nice because it supplies us with a continuous line of numbers, but there are still some problems. For example, not all polynomials with real coefficients have real-number solutions (for example, $x^2 + 1 = 0$). The set of complex numbers was created in order to resolve this inadequacy of $\mathbb{R}$. If we consider a quadratic $ax^2 + bx + c = 0$ with real coefficients, we know that our two solutions are

$$x = \frac{-b \pm \sqrt{b^2 - 4ac}}{2a}.$$

If $b^2 - 4ac \geq 0$, then we have one or two real solutions. If $b^2 - 4ac < 0$, then we have solutions

$$z_1 = -\frac{b}{2a} + \frac{\sqrt{4ac - b^2}}{2a}\sqrt{-1}$$

and

$$z_2 = -\frac{b}{2a} - \frac{\sqrt{4ac - b^2}}{2a}\sqrt{-1}.$$

These are both of form $u + v\sqrt{-1}$ where u and v are real numbers. If we *define*

$$i = \sqrt{-1}$$

we are thus led to consider the set of all numbers of form $u + iv$, with u and v real. Numbers of this form are called *complex numbers*. Since x and y are often used to denote generic real numbers, z and w are often used to denote arbitrary complex numbers. If $u, v, x,$ and y are real, then we easily see that

- the sum of two complex numbers is complex:

$$(u + iv) + (x + iy) = (u + x) + i(v + y);$$

- there is an additive identity:

$$(u + iv) + (0 + i0) = u + iv;$$

- there exist additive inverses:

$$(u + iv) + ((-u) + i(-v)) = 0 + i0;$$

- the product of two complex numbers is complex:

$$(u + iv) \cdot (x + iy) = (ux - vy) + i(yu + xv);$$

- there is a multiplicative identity:

$$(u + iv) \cdot (1 + i0) = u + iv.$$

What about multiplicative inverses? Can we divide two complex numbers? If we observe that

$$\frac{1}{u + iv} = \frac{1}{u + iv} \cdot \frac{u - iv}{u - iv} = \frac{u - iv}{u^2 + v^2}$$

$$= \frac{u}{u^2 + v^2} + i\frac{-v}{u^2 + v^2},$$

we are led to try

$$\frac{u}{u^2 + v^2} + i\frac{-v}{u^2 + v^2}$$

as the multiplicative inverse w^{-1} of $w = u + iv \neq 0 + i0$. We then define division by

$$\frac{z}{w} = z \cdot w^{-1}.$$

The complex numbers, as do the real numbers, form a field. However, and interestingly, they cannot be an ordered field (to show this is left as Exercise A.6).

The complex numbers can be realized geometrically by associating to $x + iy$ the ordered pair (x, y). This identifies $\mathbb{C}$ with $\mathbb{R}^2$. Complex numbers are thus added like vectors, via the parallelogram equality. Multiplication of complex numbers becomes clear, geometrically, if we use polar coordinates. If

$$z = r\cos\theta + ir\sin\theta, \ r \geq 0, \ 0 \leq \theta < 2\pi,$$

and

$$w = \rho \cos\phi + i\rho \sin\phi, \ \rho \geq 0, \ 0 \leq \phi < 2\pi,$$

then, as the reader should check,

$$zw = r\rho \cos(\theta + \phi) + ir\rho \sin(\theta + \phi).$$

If we think of w as fixed, then multiplication by z corresponds to rotating by the angle θ and stretching by the factor r. With $z = x + iy = r\cos\theta + ir\sin\theta$, r is called the *magnitude* of z and is denoted by $|z|$, and θ is called the *argument* of z and is denoted by $\arg(z)$. Thus, when multiplying two complex numbers, their magnitudes multiply, and their arguments add. Note that $|z| = \sqrt{x^2 + y^2}$, which is the distance from z to the origin (here we are assuming that r is nonnegative; the reader should be mindful of the usual "uniqueness-of-representation" problems associated with polar coordinates). The *real part* of z is x and will be denoted by $\mathrm{re}(z)$, and the *imaginary part* of z is y and will be denoted by $\mathrm{im}(z)$. The *complex conjugate* of z, denoted by $\bar{z}$, is $x - iy$. Observe that the solutions z_1 and z_2 of the quadratic equation given at the beginning of the section are complex conjugates.

Finally, we want to make sense out of e^z for complex values of z. This can be done in a number of equivalent ways. We want the usual rules of exponents to hold. In particular, we want to demand that

$$e^{x+iy} = e^x e^{iy}.$$

Since e^x is already defined (since x is real), we need define e^{iy} only for real values of y. We define

$$e^{iy} = \cos y + i \sin y.$$

We now make our definition of the complex exponential. For $z = x + iy$, let

$$e^z = e^x e^{iy} = e^x(\cos y + i \sin y) = e^x \cos y + ie^x \sin y.$$

In other words, e^z is the complex number with real part $e^x \cos y$ and imaginary part $e^x \sin y$.

Exercises for Appendix A

A.1 (a) Rewrite $\dfrac{2 - 5i}{1 + i}$ in the form $x + iy$.

(b) Show that $\dfrac{1}{i} = -i$.

A.2 Prove each statement below for arbitrary complex numbers w and z.

(a) $\overline{z + w} = \bar{z} + \bar{w}$.
(b) $\overline{z \cdot w} = \bar{z} \cdot \bar{w}$.
(c) $|z \cdot w| = |z| \cdot |w|$.
(d) $|z|^2 = z \cdot \bar{z}$.

(e) $|z_1 + z_2 + \cdots + z_n| \leq |z_1| + |z_2| + \cdots + |z_n|$, n any positive integer.

(f) $\max(|x|, |y|) \leq |z| \leq \sqrt{2} \cdot \max(|x|, |y|)$, where $z = x + iy$. Observe that this implies that $|re(z)| \leq |z|$ and $|im(z)| \leq |z|$.

(g) $re(z) = \frac{z + \bar{z}}{2}$ and $im(z) = \frac{z - \bar{z}}{2i}$.

A.3 Prove each statement below for arbitrary complex numbers w and z.

(a) $e^{z+w} = e^z e^w$.

(b) $e^z \neq 0$ for each complex number z.

(c) $|e^{i\theta}| = 1$ for every real number θ.

(d) $(\cos\theta + i\sin\theta)^n = \cos n\theta + i\sin n\theta$ for every real number θ. (In this exercise the value for n is intentionally left vague. For what values of n can you prove the statement? n a positive integer? n any integer? n any real number?)

(e) Does $e^z = e^w$ imply that $z = w$? Explain. (Compare this to the real case.)

A.4 Find the real and imaginary parts of $\dfrac{(\sqrt{3} + i)^6}{(1 - i)^{10}}$.

A.5 Is it true that $re(z \cdot w) = re(z) \cdot re(w)$? Either prove that it is true, or give a counterexample to show that it does not always hold.

A.6 The complex numbers cannot be ordered to give an ordered field. (This may require you looking up the definition of a "field" as well as that of an "ordered field.")

Appendix B
Basic Set Theory

In this appendix we give several definitions having to do with sets that are used throughout the main body of the text.

Throughout the text, $\emptyset$ denotes the empty set: the set with no members. We write "$x \in A$" to designate that x is a member, or element, of the set A. Likewise, we use the symbol "$\notin$" to designate nonmembership.

If A and B are two sets, the *union* of A and B is the set

$$A \cup B = \{x \mid x \in A \text{ or } x \in B\},$$

and the *intersection* of A and B is the set

$$A \cap B = \{x \mid x \in A \text{ and } x \in B\}.$$

The sets are *disjoint* if $A \cap B = \emptyset$. The set A is a *subset* of B, denoted by $A \subseteq B$ (or $A \subset B$ to denote proper inclusion), if $x \in A$ implies $x \in B$. The *complement* of A in B, $B \setminus A$, is the set of all x that are in B but are not in A (note that A need not be a subset of B for this definition). Often, A^c is used in place of $B \setminus A$ if the set B is understood. Note that $A^{cc} = A$ for each $A \subseteq \mathbb{R}^n$, that $(\mathbb{R}^n)^c = \emptyset$, and that $\emptyset^c = \mathbb{R}^n$.

We often want to discuss arbitrary collections of sets. Let $\mathcal{I}$ denote an arbitrary set such that for each $i \in \mathcal{I}$ we have a set A_i. It is important to recognize that $\mathcal{I}$ can be finite, countably infinite, or even uncountably infinite. The set $\mathcal{I}$ is then called an *index set* for the collection of sets $\{A_i\}_{i \in \mathcal{I}}$. The *union* of these sets is

$$\bigcup_{i \in \mathcal{I}} A_i = \{x \mid x \in A_i \text{ for some } i \in \mathcal{I}\},$$

and the *intersection* is the set

$$\bigcap_{i \in \mathcal{I}} A_i = \{x \mid x \in A_i \text{ for every } i \in \mathcal{I}\}.$$

The collection $\{A_i\}_{i \in \mathcal{I}}$ is a *disjoint* collection, or *pairwise disjoint* collection, if $i \neq j$ implies $A_i \cap A_j = \emptyset$.

The collection of all subsets of a given set A is denoted by 2^A.

Throughout the text we have made frequent use of the following result.

Theorem (De Morgan's Laws).   *For a collection $\{A_i\}_{i \in \mathcal{I}}$ of subsets of a set A, the following identities hold:*

$$\left(\bigcup_{i \in \mathcal{I}} A_i\right)^c = \bigcap_{i \in \mathcal{I}} A_i^c \quad \text{and} \quad \left(\bigcap_{i \in \mathcal{I}} A_i\right)^c = \bigcup_{i \in \mathcal{I}} A_i^c.$$

Two sets A and B are said to be *equivalent* if there exists a one-to-one and onto function from A to B. As usual, $\mathbb{N} = \{1, 2, 3, \ldots\}$. If a set A is equivalent to a subset $\{1, 2, \ldots, n\}$ of $\mathbb{N}$, then A is said to be *finite*. If A is not finite, then it is *infinite*. Infinite sets are either "countable" or "uncountable." *Countable* sets are sets that are equivalent to $\mathbb{N}$. An infinite set that is not countable is an *uncountable* set.

Exercises for Appendix B

B.1 Prove De Morgan's laws.

B.2 Show that the set of rational numbers $\mathbb{Q}$ is countable.

B.3 Prove that the set $\bigcup_{i \in \mathcal{I}} A_i$ is countable whenever each set A_i is countable and the index set $\mathcal{I}$ is countable.

B.4 Prove that $\mathbb{R}$ is uncountable. This exercise will probably be difficult if this material is truly new to you; you may want to ask your professor for advice.

References

[1] Adams, Malcolm and Victor Guillemin, *Measure Theory and Probability*, Wadsworth and Brooks/Cole, 1986.

[2] Adler, C.L. and James Tanton, π is the minimum value for pi, *College Math. J.* **31**, 2000, 102–106.

[3] Aliprantis, C.D. and O. Burkinshaw, *Principles of Real Analysis*, North Holland, 1981.

[4] Alvarez, S.A., L^p arithmetic, *Amer. Math. Monthly* **99**, 1992, 656–662.

[5] Aronszajn, N. and K. Smith, Invariant subspaces of completely continuous operators, *Ann. Math.* **60**, 1954, 345–250.

[6] Ash, Robert B., *Real Analysis and Probability*, Academic Press, 1972.

[7] Asplund, Edgar and Lutz Bungart, *A First Course in Integration*, Holt, Rinehart and Winston, 1966.

[8] Axler, Sheldon, Down with determinants!, *Amer. Math. Monthly* **102**, 1995, 139–154.

[9] Baire, René, Sur les fonctions de variables réelles, *Annali di Matematica pura ed applicata* **3**, 1899, 1–122.

[10] Banach, Stefan, Sur les opérations dans les ensembles abstraits et leur application aux équations intégrales, *Fundamenta Mathematicae* **3**, 1922, 133–181.

[11] Banach, Stefan, *Théorie des Opérations Linéaires*, Monografje Matematyczne, 1932.

[12] Banach, Stefan and A. Tarski, Sur la décomposition des ensembles de points en parties respectivement congruents, *Fund. Math.* **6**, 1924, 244–277.

[13] Bartle, R.G., Return to the Riemann integral, *Amer. Math. Monthly* **103**, 1996, 625–632.

[14] Bartle, R.G., *The Elements of Integration*, John Wiley and Sons, 1966.

[15] Beauzamy, B., *Introduction to Operator Theory and Invariant Subspaces*, North-Holland, 1988.

[16] Bernkopf, Michael, The development of function spaces with particular reference to their origins in integral equation theory, *Arch. History Exact Sci.* **3**, 1966, 1–96.

[17] Bernkopf, Michael, A history of infinite matrices, *Arch. History Exact Sci.* **4**, 1967/1968, 308–358.

[18] Bernstein, Sergei, Démonstration du théorème de Weierstrass, fondée sur le calcul des probabilités, *Commun. Soc. Math. Kharkov* **2** (13), 1912–1913, 1–2.

[19] Bishop, Errett, A generalization of the Stone–Weierstrass theorem, *Pacific J. Math.* **11**, 1961, 777–783.

[20] Bohnenblust, H.F. and A. Sobczyk, Extensions of functionals on complete linear spaces, *Bull. Amer. Math. Soc.* **44**, 1938, 91–93.

[21] Bollobás, Béla, *Linear Analysis*, Cambridge (second edition), 1999.

[22] Border, Kim C., *Fixed Point Theorems with Applications to Economics and Game Theory*, Cambridge, 1985.

[23] Bottazzini, Umberto, *The Higher Calculus: A History of Real and Complex Analysis from Euler to Weierstrass*, Springer-Verlag, 1986.

[24] de Branges, Louis, The Stone–Weierstrass theorem, *Proc. Amer. Math. Soc.* **10**, 1959, 822–824.

[25] Bressoud, David, *A Radical Approach to Real Analysis*, Mathematical Association of America, 1994.

[26] Brosowski and Deutsch, An elementary proof of the Stone–Weierstrass theorem, *Proc. Amer. Math. Soc.* **81**. 1981, 89–92.

[27] Brown, James Ward and Ruel V. Churchill, *Complex Variables and Applications*, McGraw-Hill (sixth edition) 1996.

[28] Burke-Hubbard, Barbara, *The World According to Wavelets*, A.K. Peters, 1996.

[29] Burkill, J.C., Henri Lebesgue, Obituary, *J. Lond. Math. Soc.* **19**, 1944, 56–64.

[30] Conway, John B., *Functions of One Complex Variable*, Springer-Verlag (second edition), 1978.

[31] Coutinho, C.S., The many avatars of a simple algebra, *Amer. Math. Monthly* **104**, 1997, 593–604.

[32] Dauben, Joseph, Georg Cantor and Pope Leo XIII: mathematics, theology, and the infinite, *Journal Hist. Ideas* **38**, 1977, 85–108.

[33] DeVito, Carl, *Functional Analysis*, Academic Press, 1978.

[34] Dieudonné, J., *History of Functional Analysis*, North-Holland, 1981.

[35] Dorier, Jean-Luc, A general outline of the genesis of vector space theory, *Hist. Math.* **22**, 1995, 227–261.

[36] Dunford, N. and J. Schwartz, *Linear Operators, Parts I, II, III*, Wiley, 1971.

[37] Dunham, William, A historical gem from Vito Volterra, *Math. Mag.* **63**, 1990, 234–236.

[38] Dydak, Jerzy and Nathan Feldman, Major theorems in compactness: a unified approach, *Amer. Math. Monthly* **99**, 1992, 220–227.

[39] Dym, H. and H.P. McKean, *Fourier Series and Integrals*, Academic Press 1972.

[40] Enflo, P., On the invariant subspace problem in Banach spaces, *Acta Math.* **158**, 1987, 213–313.

[41] Enflo, P., B. Beauzamy, E. Bombieri, and H. Montgomery, Products of polynomials in many variables, *Journal of Number Theory* **36**, 1990, 219–245.

[42] Fischer, E., Sur la convergence en moyenne, *Comptes Rendus Acad. Sci. Paris* **144**, 1907, 1022–1024.

[43] Folland, Gerald B., *Fourier Analysis and Its Applications*, Wiley, (second edition), 1999.

[44] Folland, Gerald B., *Real Analysis*, Wadsworth and Brooks/Cole, 1992.

[45] Folland, Gerald B. and Alladi Sitaram, The uncertainty principle: a mathematical survey, *J. Four. Ana. Appl.* **3**, 1997, 207–238.

[46] Frazier, Michael and R. Meyer-Spasche, *An Introduction to Wavelets Through Linear Algebra*, Springer-Verlag, 1999.

[47] Friedman, Avner, *Foundations of Modern Analysis*, Holt, Rinehart and Winston, 1970.

[48] Gamelin, T.W., *Uniform Algebras*, Prentice-Hall, 1969.

[49] González-Velasco, Enrique A., *Fourier Analysis and Boundary Value Problems*, Academic Press, 1995.

[50] Grabiner, S., A short proof of Runge's theorem, *Amer. Math. Monthly* **83**, 1976, 807–808.

[51] Grattan-Guiness, Ivor, *Joseph Fourier 1768-1830*, M.I.T. Press, 1972.

[52] Grattan-Guiness, Ivor, A sideways look at Hilbert's twenty-three problems of 1900, *Amer. Math. Soc. Not.* **40** (4), 2000, 752–757.

[53] Gzyl, H. and J.L. Palacios, The Weierstrass approximation theorem and large deviations, *Amer. Math. Monthly* **104**, 1997, 650–653.

[54] Hairer, E. and G. Wanner, *Analysis by its History*, Springer-Verlag, 1996.

[55] Halmos, Paul R., What does the spectral theorem say?, *Amer. Math. Monthly* **70**, 1963, 241–247.

[56] Halmos, Paul R., Ten problems in Hilbert space, *Bull. Amer. Math. Soc.* **76**, 1970, 887–993.

[57] Halmos, P.R., The legend of John von Neumann, *Amer. Math. Monthly* **80**, 1973, 382–394.

[58] Halmos, Paul R., *A Hilbert Space Problem Book*, Springer-Verlag, 1974.

[59] Halmos, Paul R., *Naive Set Theory*, Springer-Verlag, 1982.

[60] Haunsperger, Deanna and Steve Kennedy, Coal miner's daughter, *Math. Hor.*, April 2000.

[61] Hawkins, Thomas, *Lebesgue's Theory of Integration: Its Origins and Developments*, Univ. of Wisconsin, 1970.

[62] Helly, Eduard, Über lineare Funktionaloperationen, *Sitzber. Akad. Wiss. Wien* **121**, 1912, 265–297.

[63] Heuser, H.G., *Functional Analysis*, Wiley, 1982.

[64] Hewitt, E., The role of compactness in analysis, *Amer. Math. Monthly* **67**, 1960, 499–516.

[65] Hochstadt, Harry, Eduard Helly, Father of the Hahn-Banach theorem, *Math. Intelligencer* **2**, 1980, 123–125.

[66] Hoffman, Kenneth, *Banach Spaces of Analytic Functions*, Prentice-Hall, 1962.

[67] Hollis, Selwyn, Cones, k-cycles of linear operators, and problem B4 on the 1993 Putnam competition, *Math. Mag.* **72**, 1999, 299–303.

[68] Jacod, Jean and Philip Protter, *Probability Essentials*, Springer-Verlag, 2000.

[69] Jauch, Josef, *Foundations of Quantum Mechanics*, Addison-Wesley, 1968.

[70] Johnsonbaugh, Richard and W.E. Pfaffenberger, *Foundations of Mathematical Analysis*, Marcel Dekker, 1981.

[71] Jones, Frank, *Lebesgue Integration on Euclidean Space*, Jones and Bartlett, 1993.

[72] Kahane, Jean-Pierre, A century of interplay between Taylor series, Fourier series and Brownian motion, *Bull. London Math. Soc.* **29**, 1997, no. 3, 257–279.

[73] Kałuża, Roman, *Through a Reporter's Eyes: The Life of Stefan Banach*, Birkhäuser, 1996.

[74] Katz, Irving J., An inequality of orthogonal complements, *Math. Mag.* **65**, 1992, 258–259.

[75] Kendall, David, Obituary of Maurice Fréchet, *J. Roy. Statist. Soc. Ser. A* **140** (4), 1977, 566.

[76] Kline, Morris, *Mathematical Thought from Ancient to Modern Times*, Oxford, 1972.

[77] Körner, T.W., *Fourier Analysis*, Cambridge, 1988.

[78] Kraft, Roger, What's the difference between Cantor sets?, *Amer. Math. Monthly* **101**, 1994, 640–650.

[79] Kuratowwski, Kazimierz, *A Half Century of Polish Mathematics*, Pergamon, 1973.

[80] Lebesgue, Henri, *Intégrale, Longuere, Aire*, Ph.D. thesis, University of Nancy, 1902.

[81] Lebesgue, Henri, Sur l'existence des derivées, *Comptes Rendus* **136**, 1903, 659–661.

[82] Lebesgue, Henri, *Measure and Integral*, edited with a biographical essay by K.O. May, Holden-Day, 1966.

[83] Lehto, Olli, *A History of the International Mathematical Union*, Springer-Verlag, 1998.

[84] Lomonosov, Victor, Invariant subspaces for operators commuting with a compact operator, *Funct. Anal. i Prilozen* **7**, 1973, 55–56.

[85] Machado, Silvio, On Bishop's generalization of the Stone–Weierstrass theorem, *Indag. Math.* **39**, 1977, 218–224.

[86] Mackey, G.W., Marshall Harvey Stone, Obituary, *Amer. Math. Soc. Not.* **36**, 1989, 221–223.

[87] Macrae, Norman, *John von Neumann: The Scientific Genius who Pioneered the Modern Computer, Game Theory, Nuclear Deterrence, and Much More*, American Mathematical Society, 1992.

[88] Marsden, Jerrold and Michael Hoffman, *Basic Complex Analysis*, Freeman (second edition), 1993.

[89] Marsden, Jerrold and Michael Hoffman, *Elementary Classical Analysis*, Freeman (third edition), 1999.

[90] McDonald, John N. and N.A. Weiss, *A Course in Real Analysis*, Academic Press, 1999.

[91] Miamee, A.G., The inclusion $L^p(\mu) \subset L^q(\nu)$, *Amer. Math. Monthly* **98**, 1991, 342–345.

[92] Moe, Karine and Karen Saxe, *Fixed Point Theorems in Economics*, COMAP, 2000.

[93] Monna, A.F., The concept of function in the nineteenth and twentieth centuries, in particular with regard to the discussions between Baire, Borel, and Lebesgue, *Arch. History Exact Sci.* **9**, 1972, 57–84.

[94] Monna, A.F., *Functional Analysis in Historical Perspective*, Halsted, 1973.

[95] Mycielski, Jan, Two constructions of Lebesgue's measure, *Amer. Math. Monthly* **85**, 1978, 257–259.

[96] Narici, Lawrence and Edward Beckenstein, The Hahn–Banach theorem: the life and times, *Topology Appl.* **77**, 1997, 193–211.

[97] von Neumann, John, Zur Theorie des Gesellschaftsspiele, *Math. Annalen* **100**, 1928, 295–320.

[98] von Neumann, John, Allgemeine Eigenwerttheorie Hermitescher Funktionaloperatoren, *Math. Annalen* **102**, 1929/1930, 49–131.

[99] von Neumann, John, *Mathematical Foundations of Quantum Mechanics*, Princeton, 1955.

[100] Pfeffer, Washek F., *Integrals and Measures*, Marcel Dekker, 1977.

[101] Radjavi, H. and P. Rosenthal, The invariant subspace problem, *Math. Intell.* **4**, 1982 no. 1, 33–37.

[102] Ransford, Thomas, A short elementary proof of the Bishop–Stone–Weierstrass theorem, *Math. Proc. Camb. Phil. Soc.* **96**, 1984, 309–311.

[103] Read, Charles, Quasinilpotent operators and the invariant subspace problem, *J. London Math. Soc.* (2) **56**, 1997 no. 3, 595–606.

[104] Reid, Constance, *Hilbert*, Springer-Verlag, 1970.

[105] Riesz, Frigyes, Untersuchungen über Systeme integrierbarer Funktionen, *Math. Ann.* **69**, 1910, 449–497.

[106] Riesz, Frigyes, Sur les opérations dans les ensembles abstraits et leur application aux équations intégrales, *Fundamenta Mathematicae* **3**, 1922, 133–181.

[107] Riesz, Frigyes and Béla Sz.-Nagy, *Functional Analysis*, Frederick Ungar, 1955.

[108] Roginski, W.W., Frederic Riesz, Obituary, *J. Lond. Math. Soc.* **31**, 1956, 508–512.

[109] Ross, Kenneth A., *Elementary Analysis: The Theory of Calculus*, Springer-Verlag, 1980.

[110] Rudin, Walter, *Principles of Mathematical Analysis*, McGraw-Hill, 1953.

[111] Rudin, Walter, *Real and Complex Analysis*, McGraw-Hill, (second edition) 1974.

[112] Rudin, Walter, *Functional Analysis*, McGraw-Hill, (second edition) 1991.

[113] Schechter, Martin, *Operator Methods in Quantum Mechanics*, North Holland, 1981.

[114] Smithies, F., John von Neumann, Obituary, *J. Lond. Math. Soc.* **34**, 1959, 373–374.

[115] Solovay, Robert, A model of set theory in which every set of reals is Lebesgue measurable, *Ann. Math.* **92**, 1970, 1–56.

[116] Steinhaus, Hugo, Additive und stetige Funktionaloperationen, *Math. Zeitschrift* **5**, 1919, 186–221.

[117] Steen, Lynn, Highlights in the history of spectral theory, *Amer. Math. Monthly* **80**, 1973, 359–381.

[118] Stone, Marshall H., Applications of the theory of Boolean rings to general topology, *Trans. Amer. Math. Soc.* **41**, 1937, 375–481.

[119] Stone, Marshall H., The generalized Weierstrass approximation theorem, *Math. Mag.* **21**, 1948, 167–184 and 237–254.

[120] Stone, Marshall H., Mathematics in continental China, 1949–1960, *Amer. Math. Soc. Not.* **8**, 1961, 209–215.

[121] Strang, Gilbert and Truong Nguyen, *Wavelets and Filter Banks*, Wellesley Cambridge, 1996.

[122] Stroock, Daniel W., *A Concise Introduction to the Theory of Integration*, Birkhäuser (second edition), 1994.

[123] Taylor, Angus E., A study of Maurice Fréchet: I. His early work on point set theory and the theory of functionals, *Arch. Hist. Exact Sci.* **27**, 1982, 233–295.

[124] Taylor, Angus and David Lay, *Introduction to Functional Analysis*, Wiley (second edition) 1980.

[125] Thorsen, Bobette, The eigenvalues of an infinite matrix, *College Math. J.* **31**, 2000, 107–110.

[126] Vitali, Giuseppe, *Sul problema della misura dei gruppi de punti di una retta*, Mem. Accad. de Bologna, 1905.

[127] Weir, Alan J., *Lebesgue Integration and Measure*, Cambridge, 1973.

[128] Wheeden R. and A. Zygmund, *Measure and Integral: An Introduction to Real Analysis*, Marcel Dekker, 1977.

[129] Young, N., *An Introduction to Hilbert Space*, Cambridge, 1988.

[130] Zaanen, A.C., Continuity of measurable functions, *Amer. Math. Monthly* **93**, 1986, 128–130.

Index

194   Index

Cauchy, Augustin Louis, 7, 32
Cauchy–Schwarz Inequality, 7
Cesàro operator, 129
characteristic function, 45
clopen set, 29
closed set in a metric space, 16
closure of a set, 16
compact linear operator, 107
compact set, 16
compact support, function with, 73
complement of a set, 185
complete metric space, 23
complete orthonormal sequence, 77
complete subset in a metric space, 23
complex conjugate of a complex number,
    183
composition operator, 129
connected set, 158
continuous function on a metric space, 5
contraction mapping, 161
convergence in a metric space, 5
convergence in mean, 86
convolution, 132
countable set, 186
countably additive set function, 36
countably subadditive set function, 38
counting measure, 58
cover of sets, 16

De Branges, Louis, 144
De Morgan's laws, 186
dense set, 16
diagonal operator, 102
    characterization of compactness of,
        108
Dirac, Paul, 167
discrete metric, 4
discrete probability space, 34
disjoint sets, 185
dual space, 154
Dunham, William, 150

E°, 16
Einstein, Albert, 27
Enflo, Per, 110, 112, 122
equicontinuous function, 19
equivalent norms, 30
equivalent sets, 186
essentially bounded function, 61

Euclidean norm, 5
Eudoxos, 32

f_-, 45
f_+, 45
Fatou's lemma, 52, 64
Fatou, Pierre, 52
Fields Medals, 26
finite-dimensional, 11
finite-rank operator, 110
finitely subadditive set function, 69
first category set, 147
Fourier coefficients, 82
Fourier series, 82
Fourier transform, 167
Fourier, Joseph, 4, 75, 79
Fréchet, Maurice, 1, 11, 33
Fredholm operator of the first kind, 94
Fredholm operator of the second kind, 94
Fredholm, Erik Ivar, 12, 68, 94
fundamental theorem of algebra, 104
fundamental theorem of calculus, 56

Grabiner, Sandy, 145
Gram–Schmidt process, 77

Haar functions, 78
Haar, Alfréd, 79
Hadamard, Jacques, 12
Hahn, Hans, 151
Hahn–Banach theorem (complex case),
    156
Hahn–Banach theorem (real case), 155
Hamel basis, 11
Hausdorff, Felix, 13
Heine, Eduard, 18
Heine–Borel theorem, 17, 18
Heisenberg, Werner, 167
Helly, Eduard, 154
Hermite functions, 78
Hermite, Charles, 78, 114
Hermitian operator, 114
Hermitian symmetry property of an inner
    product, 6
Hilbert space, 23
Hilbert's 23 problems, 26
Hilbert, David, 1, 24, 33, 128
Hilden, Hugh, 120
Hölder conjugate, 60

Undergraduate Texts in Mathematics

(continued from page ii)